Lecture Notes in Physics

Volume 887

For further volumes:
http://www.springer.com/series/5304

The Lecture Notes in Physics

The series Lecture Notes in Physics (LNP), founded in 1969, reports new developments in physics research and teaching-quickly and informally, but with a high quality and the explicit aim to summarize and communicate current knowledge in an accessible way. Books published in this series are conceived as bridging material between advanced graduate textbooks and the forefront of research and to serve three purposes:

- to be a compact and modern up-to-date source of reference on a well-defined topic
- to serve as an accessible introduction to the field to postgraduate students and nonspecialist researchers from related areas
- to be a source of advanced teaching material for specialized seminars, courses and schools

Both monographs and multi-author volumes will be considered for publication. Edited volumes should, however, consist of a very limited number of contributions only. Proceedings will not be considered for LNP.

Volumes published in LNP are disseminated both in print and in electronic formats, the electronic archive being available at springerlink.com. The series content is indexed, abstracted and referenced by many abstracting and information services, bibliographic networks, subscription agencies, library networks, and consortia.

Proposals should be sent to a member of the Editorial Board, or directly to the managing editor at Springer:

Christian Caron
Springer Heidelberg
Physics Editorial Department I
Tiergartenstrasse 17
69121 Heidelberg/Germany
christian.caron@springer.com

Walter Schirmacher

Theory of Liquids and Other Disordered Media

A Short Introduction

 Springer

Walter Schirmacher
Institut für Physik Komet 331
Universität Mainz
Mainz
Germany

ISSN 0075-8450   ISSN 1616-6361 (electronic)
Lecture Notes in Physics
ISBN 978-3-319-06949-4   ISBN 978-3-319-06950-0 (eBook)
DOI 10.1007/978-3-319-06950-0
Springer Cham Heidelberg New York Dordrecht London

Library of Congress Control Number: 2014944752

Printed on acid-free paper

Springer is part of Springer Science+Business Media (www.springer.com)

For Linda

"What is the use of a book without pictures or conversations?"
Lewis Carroll, Alice's Adventures in Wonderland

Preface

The physics of liquids, solutions, glasses, and macromolecular materials comprises a very large area of physical research. Nowadays these fields grow together and form the rapidly expanding field of *soft-matter physics*, which also includes the investigation of colloids, liquid crystals, and biological materials. Within the large scientific community of soft-matter research the liquid and/or macromolecular materials are investigated mostly with physicochemical experimental techniques and by means of computer simulations. The latter method has in the last 50 years grown from a part of theoretical physics into what is now called *computer experiment*, because the simulation data are analyzed in the same fashion as experimental data. The only difference between simulational and experimental data is that the underlying equation of motion (Newton's or Schrödinger's equation) is known, but that is in principle also the case for the experiment. One advantage of the simulation is that model systems can be studied, which are simpler than the real materials. Another is that microscopic information can be obtained, which is out of scope of experiments.

However, a theoretical understanding of the behavior of complex materials requires more than the knowledge of the underlying microscopic equations of motions. Therefore theoretical concepts on a macroscopic level are required. A number of such concepts are presented in these lecture notes.

The present lecture notes arose from two courses: "Theory of Liquids and Polymers I and II," held at the institute for functional materials (Prof. W. Petry and Prof. P. Müller-Buschbaum, E13) at the Physics Department of Technische Universität München in the academic year 2006/2007. The aim of these courses was to create a mutual understanding and a common theoretical language among students and scientists working in this institute on such different subjects as liquid metals, glasses, polymers, and biological materials. The handouts presented at these courses formed the backbone of the present lecture notes.

The division of the notes into structure/thermodynamics and dynamics arises from the fact that in classical systems (i.e., materials in which quantum effects are not dominant) the structural and thermal properties can be studied without the knowledge of the dynamics. On the other hand, the dynamics of a liquid or other soft-matter material is strongly dependent on its structure.

I have profited very much from illuminating discussions concerning the subject-matter of the present lecture notes with (in alphabetic order) Kurt Binder, Jürgen Bosse, Robert Evans, Thomas Franosch, Matthias Fuchs, Helmut Gabriel, Wolfgang Götze, Jürgen Horbach, Andreas Meyer, Bijan Movaghar, Peter Müller-Buschbaum, Philippe Nozières, Winfried Petry, Giancarlo Ruocco, Henner Ruppersberg, Friederike Schmid, Rolf Schilling, Alfons Schulte, Francesco Sciortino, Tullio Scopigno, Harald Sinn, Matthias Sperl, Gabriele Viliani, Thomas Voigtmann, and Joachim Wuttke.

I am indebted to Christian Caron and to the editorial board of LNP of Springer-Verlag for carefully reading the manuscript and help for improving the manuscript.

Mainz   Walter Schirmacher
July 25, 2014

Contents

Chapter 1
Introduction

From the standpoint of a crystallographer the term structure of a liquid is ill-defined, as the crystal structure means the arrangement of the unit cell of a crystal, which is then repeated by the lattice translations. However the term structure has become common for liquid and amorphous materials describing the statistics of the interatomic distances. As we shall see, this statistics can be measured by X-ray (and neutron) diffraction as the crystalline structure.

We shall not give an introduction to the general statistical mechanics of an interacting many-body system but take the knowledge of these concepts for granted.

For a thorough introduction to the statistical mechanics of liquids I refer to the standard textbooks by Boon and Yip [2], Balucani and Zoppi [1], Egelstaff [4], Hansen and McDonald [9], and March and Tosi [11]. Standard textbooks for Polymer science are e.g., de Gennes [3] and Sperling [13].

Our starting point is the description of distribution functions which describe the statistical arrangements of atoms or molecule in a simple liquid.[1]

The main concepts will involve *mean field theories* like the Perkus–Yevick theory and the Random Phase Approximation which relate the forces to the distribution functions.

The Random Phase Approximation (RPA) will also turn out to be the basis of the thermodynamics of *binary solutions*, which has been used by Flory [5] and Huggins [10] to discuss polymer melts and solutions, as we shall do in the end of Part I.

Before turning to the discussion of polymer melts and solutions we shall try to get acquainted to *scaling concepts* by discussing *random walks* and *fractals*. On this basis we then shall discuss the scaling concepts of polymer science. Because the concept of a random walk is needed to discuss the statistics of a polymer chain, the equation of motion for the statistics of a Brownian particle (diffusion equation)

---

[1]We are going to call a liquid *simple* if it can be described in terms of a set of *pairwise intermolecular forces or potentials*.

W. Schirmacher, *Theory of Liquids and Other Disordered Media*, Lecture Notes in Physics 887, DOI 10.1007/978-3-319-06950-0_1,
© Springer International Publishing Switzerland 2015

is already introduced and discussed in this context and not in the second part of the present lectures, which is devoted to dynamics.

Part II of the lecture series is then entirely devoted to the discussion of the dynamics of simple and complex liquids. A useful concept for such a discussion has proved to be the generalized hydrodynamics introduced by Mori [12] and Zwanzig [14] (see also Zwanzig [15] and Forster [6]). It will be demonstrated, that another mean-field-like theory, the mode-coupling theory [7, 8] is capable to both describe the salient features of the collective dynamics of a simple liquid (Chap. 8) and the structural arrest with increasing density and/or decreasing temperature (*glass transition*, Chap. 11). In Chaps. 9 and 10 different aspects of diffusive single-particle motion and polymer dynamics are discussed.

References

1. Balucani, U., Zoppi, M.: Dynamics of the Liquid State. Clarendon Press, Oxford (1983)
2. Boon, J.P., Yip, S.: Molecular Hydrodynamics. McGraw-Hill, New York (1980)
3. de Gennes, P.G.: Scaling Concepts in Polymer Physics. Cornell University Press, Ithaka (1979)
4. Egelstaff, P.: An Introduction to the Liquid State. Academic, New York (1967)
5. Flory, P.J.: Chem. Phys. **9**, 660 (1941)
6. Forster, D.: Hydrodynamic Fluctuations, Broken Symmetry and Correlation Functions. Benjamin, Reading (1975)
7. Götze, W.: In: Hansen, J.P., Levesque, D., Zinn-Justin, J. (eds.) Liquids, Freezing and the Glass Transition. Elsevier, Amsterdam (1991)
8. Götze, W.: Complex Dynamics of Glass-Forming Liquids: A Mode-Coupling Theory. Oxford University Press, Oxford (2008)
9. Hansen, J.-P., McDonald, I.: Theory of Simple Liquids. Academic, New York (1986)
10. Huggins, H.L.: Chem. Phys. **9**, 440 (1941)
11. March, N., Tosi, M.: Atomic Dynamics in Liquids. Dover, New York (1991)
12. Mori, H.: Prog. Theor. Phys. **33**, 423 (1965)
13. Sperling, L.H.: Introduction to Physical Polymer Science. Wiley, Weinheim (2006)
14. Zwanzig, R.: Phys. Rev. **124**, 983 (1961)
15. Zwanzig, R.: Nonequilibrium Statistical Mechanics. Oxford University Press, Oxford (2001)

Part I
Structure and Thermodynamics

Chapter 2
Structure of Liquids

2.1 Molecular Distribution Functions

The *structure* of a liquid is governed by the statistical distribution of the centers of gravity of the atoms or molecules. Of course the latter keep moving, but we can ask about the atomic distributions if one could perform a snapshot of the atomic arrangements. The average statistics of such snapshots is what we call the (static) *structure* of the liquid.

We therefore pose the question of how a collection of $N \approx 10^{23}$ atoms or molecules (or much less in a computer simulation) are distributed inside a certain volume V. The probability for these particles to occupy volume elements $d^3\mathbf{r}_1, d^3\mathbf{r}_2, \ldots d^3\mathbf{r}_N$ around positions $\mathbf{r}_1, \mathbf{r}_2, \ldots \mathbf{r}_N$ is given by

$$P(\mathbf{r}_1, \mathbf{r}_2, \ldots \mathbf{r}_N) d^3\mathbf{r}_1, d^3\mathbf{r}_2, \ldots d^3\mathbf{r}_N . \tag{2.1}$$

$P(\mathbf{r}_1, \mathbf{r}_2, \ldots \mathbf{r}_N)$ is the *probability density* of the configuration $\{\mathbf{r}_1, \mathbf{r}_2, \ldots \mathbf{r}_N\}$ and is normalized to 1:

$$\int_V \prod_{\alpha=1}^{N} d^3\mathbf{r}_\alpha \, P(\mathbf{r}_1, \mathbf{r}_2, \ldots \mathbf{r}_N) = 1 . \tag{2.2}$$

If a physical quantity A depends on the position of the particles the *configurational average* can be calculated as

$$\langle A \rangle = \int_V \prod_{\alpha=1}^{N} d^3\mathbf{r}_\alpha \, A(\mathbf{r}_1, \mathbf{r}_2, \ldots \mathbf{r}_N) P(\mathbf{r}_1, \mathbf{r}_2, \ldots \mathbf{r}_N) . \tag{2.3}$$

W. Schirmacher, *Theory of Liquids and Other Disordered Media*, Lecture Notes in Physics 887, DOI 10.1007/978-3-319-06950-0_2,
© Springer International Publishing Switzerland 2015

One can select $n < N$ particles in order to define the *reduced n-particle densities*

$$\rho^{(n)}(\mathbf{r}_1, \mathbf{r}_2, \ldots \mathbf{r}_n) = \frac{N!}{(N-n)!} \int_V \prod_{\alpha=n+1}^{N} d^3 r_\alpha \, P(\mathbf{r}_1, \mathbf{r}_2, \ldots \mathbf{r}_N) \,. \qquad (2.4)$$

In the case of a *complete random arrangement*, which is realized in an ideal gas, we have

$$\rho^{(n)}(\mathbf{r}_1, \mathbf{r}_2, \ldots \mathbf{r}_n) = \left(\frac{N}{V}\right)^n \equiv \rho_0^n \,. \qquad (2.5)$$

where ρ_0 is called the *homogeneous density*. The deviation from this random distribution is given by the *n-particle correlation functions* $g^{(n)}$, which are defined as follows

$$\rho^{(n)}(\mathbf{r}_1, \mathbf{r}_2, \ldots \mathbf{r}_n) = \rho_0^n g^{(n)}(\mathbf{r}_1, \mathbf{r}_2, \ldots \mathbf{r}_n) \,. \qquad (2.6)$$

In all macroscopically *homogeneous* systems, especially in simple liquids, we have

$$\rho^{(1)}(\mathbf{r}_1) = \rho_0 \qquad (2.7)$$

$$g^{(1)}(\mathbf{r}_1) = 1 \,. \qquad (2.8)$$

In a material, which is *homogeneous* and *isotropic* on a macroscopic scale we have

$$g^{(2)}(\mathbf{r}_1, \mathbf{r}_2) = g(|\mathbf{r}_1 - \mathbf{r}_2|) \qquad (2.9)$$

$g(r)$ is called *radial pair correlation function* or *radial pair distribution function* and can, as we shall see in the next section, be determined by neutron or x-ray diffraction.

$\rho_0 4\pi r^2 g(r) dr$ is the probability for a particle being present in a spherical shell of width dr around a given particle at the origin with distance r. Consequently $Z(R) = \rho_0 4\pi \int_0^R dr \, r^2 g(r)$ is the mean number of particles around a given one within a sphere of radius R. If R is chosen to be near the first minimum of $g(R)$, $Z(R)$ is the (mean) number of nearest neighbors of a given particle (mean coordination number). $g(r)$ can be represented mathematically as

$$g(r) = \frac{1}{\rho_0} \sum_{\alpha \neq 0} \langle \delta(\mathbf{r} - \mathbf{r}_\alpha) \rangle \,, \qquad (2.10)$$

where r_α is a vector from the particle 0 at the origin to another particle with label α.

Fig. 2.1 Geometry for a
scattering experiment with
incoming plane wave and
outgoing spherical wave

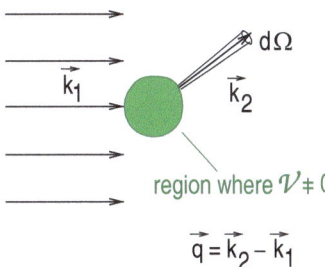

$$\vec{q} = \vec{k}_2 - \vec{k}_1$$

2.2   Scattering Theory

We want to describe the (elastic) scattering of X-rays or neutrons from a simple
liquid sample. We study an ingoing plane wave (1st term) and an outgoing scattered
spherical wave (2nd term), as depicted in Fig. 2.1, of the following asymptotic form

$$\psi(\mathbf{r}) \overset{|\mathbf{r}| \to \infty}{=} e^{i\mathbf{k}_1 \cdot \mathbf{r}} + f(\theta)\frac{1}{r}e^{ik_2 r} \tag{2.11}$$

The scattering cross-section into the solid angle element $d\Omega$ in the direction of
$\mathbf{k}_2$ is then given by the modulus-square of the scattering amplitude

$$\frac{d\sigma}{d\Omega} = |f(\theta)|^2 \tag{2.12}$$

If the scattering potential (the potential between the scattered rays and the
particles) can be decomposed as

$$\mathcal{V}(\mathbf{r}) = \sum_{\alpha=1}^{N} v(\mathbf{r} - \mathbf{r}_\alpha) \tag{2.13}$$

the scattering amplitude is given in lowest approximation

$$f(\theta) = -\frac{m}{2\pi\hbar^2}\langle \mathbf{k}_2|\mathcal{V}|\mathbf{k}_1\rangle = -\frac{m}{2\pi\hbar^2}\mathcal{V}(\mathbf{q}) = -\frac{m}{2\pi\hbar^2}\sum_{\alpha=1}^{N} e^{i\mathbf{q}\cdot\mathbf{r}_\alpha} v(\mathbf{q})$$

$$\equiv \sum_{\alpha=1}^{N} e^{i\mathbf{q}\cdot\mathbf{r}_\alpha} f(\mathbf{q}), \tag{2.14}$$

$f(\mathbf{q})$ is called *form factor* and has the unit of length. In the case of *energy unresolved*
neutron or X-ray diffraction there is no net energy exchange with the sample, i.e.,

$|\mathbf{k}_1| = |\mathbf{k}_2|$. In an isotropic material the form factor depends only on the modulus of the exchanged momentum, which is given by

$$q = |\mathbf{k}_1 - \mathbf{k}_2| = k_1 \sqrt{2[1 - \cos(\theta)]} = \frac{4\pi}{\lambda} \sin\left(\frac{\theta}{2}\right), \qquad (2.15)$$

where θ is the angle between $\mathbf{k}_1$ and $\mathbf{k}_2$ and $\lambda = \frac{2\pi}{|k_1|}$ is the wavelength.

In the case of neutrons, which scatter from the nuclei, whose potential $v(\mathbf{r})$ is extremely short-ranged, $f(q)$ does not depend on q (in the range of interest $q < 20\,\text{Å}^{-1}$) and is called *scattering length* and is denoted by the letter b.

Inserting (2.14) into (2.12) and averaging over an ensemble of different configurations we obtain

$$\frac{d\sigma}{d\Omega} = |f(q)|^2 \left\langle \sum_{\alpha,\beta=1}^{N} e^{i\mathbf{q}\cdot[\mathbf{r}_\alpha - \mathbf{r}_\beta]} \right\rangle \equiv |f(q)|^2 N S(q), \qquad (2.16)$$

where $S(q)$ is the *static structure factor*

$$S(q) = \frac{1}{N} \left\langle \sum_{\alpha,\beta=1}^{N} e^{i\mathbf{q}\cdot[\mathbf{r}_\alpha - \mathbf{r}_\beta]} \right\rangle \qquad (2.17)$$

By means of (2.17) the structure factor can be related to the radial pair distribution as follows:

$$S(q) = 1 + \frac{1}{N} \left\langle \sum_{\substack{\alpha,\beta=1 \\ \alpha \neq \beta}}^{N} e^{i\mathbf{q}\cdot[\mathbf{r}_\alpha - \mathbf{r}_\beta]} \right\rangle$$

$$= 1 + \frac{1}{N} \left\langle \int_V \mathbf{r} e^{i\mathbf{q}\mathbf{r}} \sum_{\substack{\alpha,\beta=1 \\ \alpha \neq \beta}}^{N} \delta(\mathbf{r} - \mathbf{r}_\alpha + \mathbf{r}_\beta) \right\rangle = 1 + \rho_0 \int_V d\mathbf{r} e^{i\mathbf{q}\mathbf{r}} g(r) \qquad (2.18)$$

where the last equality follows from (2.10) and recognizing that one of the summations in the double sum is redundant. A complication arises by realizing that the function $g(r)$ does not decay for large r but, instead, takes the value 1. If one substracts this value and takes the Fourier transform of 1 separately one obtains

$$S(q) = 1 + \rho_0 \int_{-\infty}^{\infty} d^3 \mathbf{r} e^{i\mathbf{q}\mathbf{r}} [g(r) - 1] + \rho_0 \delta(\mathbf{q}) \qquad (2.19)$$

As the delta function cannot be measured one usually represents the structure factor as

$$S(q) = 1 + \rho_0 \int_{-\infty}^{\infty} d^3 \mathbf{r} e^{i\mathbf{q}\mathbf{r}} [g(r) - 1] \qquad (2.20)$$

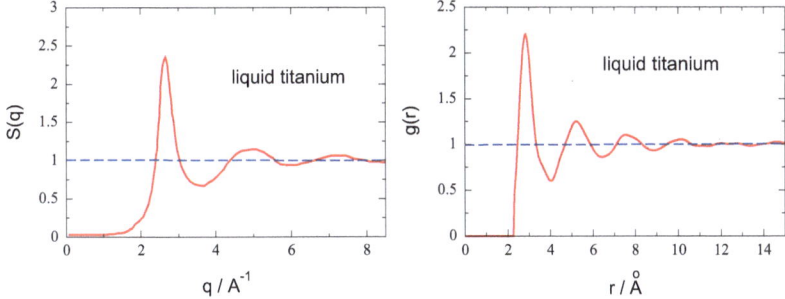

Fig. 2.2 *Left*: Structure factor of liquid titanium, measured by Lee et al. [4]. *Right*: Corresponding pair distribution function calculated via Fourier-back transform from (2.21) with $\rho_0 = 0.05$

We can now take advantage of the fact that $g(r)$ only depends on the modulus of **r** and performing the angle integral (see (A.3) of the appendix) to obtain

$$S(q) = 1 + \frac{4\pi\rho_0}{q} \int_0^\infty dr \ r \sin(qr)[g(r) - 1] \tag{2.21}$$

In Fig. 2.2 we show the example of the structure factor of liquid titanium, as measured by Lee et al. [4] by means of X-ray diffraction, together with its Fourier transform.

2.3   Thermodynamic Relations

We are now assuming that the liquid under consideration can be described by a classical Hamiltonian

$$\mathcal{H} = \sum_{\alpha=1}^N \frac{1}{2} m\dot{\mathbf{r}}_\alpha^2 + \frac{1}{2}\sum_{\alpha\neq\beta} \phi(|\mathbf{r}_\alpha - \mathbf{r}_\beta|) \ . \tag{2.22}$$

There are the following thermodynamic relationships (*equations of state*):
Potential Energy

$$\frac{1}{N}\langle E_{\text{pot}}\rangle = 2\pi\rho_0 \int_0^\infty dr \ r^2\phi(r)g(r) \tag{2.23}$$

Pressure

$$\langle P\rangle\frac{V}{Nk_BT} = \langle P\rangle\frac{1}{\rho_0 k_B T} = 1 - \frac{2\pi\rho_0}{3k_BT}\int_0^\infty dr \ r^3\phi'(r)g(r) \tag{2.24}$$

Number fluctuations and isothermal compressibility κ_T

$$\frac{\langle N \rangle^2 - \langle N^2 \rangle}{N} = \rho_0 k_B T \left(-\frac{1}{V} \frac{\partial V}{\partial P} \right)_T = \rho_0 k_B T \kappa_T$$

$$= S(q = 0) = 1 + 4\pi\rho_0 \int_0^\infty dr \, r^2 [g(r) - 1] \qquad (2.25)$$

2.4   Direct Correlation Function

The static structure factor can be split in an obvious way into a *self* and *distinct* part by separating the $\alpha \neq \beta$ term in the summation over the particles:

$$S(q) = \frac{1}{N} \left\langle \sum_{\alpha,\beta=1}^N e^{i\mathbf{q}\cdot[\mathbf{r}_\alpha - \mathbf{r}_\beta]} \right\rangle = 1 + \frac{1}{N} \left\langle \sum_{\alpha \neq \beta}^N e^{i\mathbf{q}\cdot[\mathbf{r}_\alpha - \mathbf{r}_\beta]} \right\rangle \equiv 1 + \rho_0 h(q) \quad (2.26)$$

The self part is just equal to unity, and the distinct part is ρ_0 times the Fourier transform of the *deviation of $g(r)$ from unity*, i.e.,

$$h(r) = g(r) - 1 . \qquad (2.27)$$

We now sub-divide the correlation function $h(r)$ into a part which involves only a particular pair of atoms, say $\mathbf{r}_1$ and $\mathbf{r}_2$ and a part which involves more than two atoms. Following Ornstein and Zernike, the first term is called *direct correlation function*. The second part can be generated by combining several direct functions in the following way:

$$h(r_{12}) = c(r_{12}) + \rho_0 \int d^3\mathbf{r}_3 c(r_{13})c(r_{32})$$

$$+ \rho_0^2 \int d^3\mathbf{r}_3 \int d^3\mathbf{r}_4 c(r_{13})c(r_{34})c(r_{42}) + \ldots \qquad (2.28)$$

The terms under the $\mathbf{r}_3$ integral can be summed again to give the function $h(r_{32})$:

$$h(r_{12}) = c(r_{12}) + \rho_0 \int d^3\mathbf{r}_3 c(r_{13})h(r_{32}) \qquad (2.29)$$

This is the so-called *Ornstein–Zernike equation*. It is the starting point for some *integral equation theories* for $g(r)$ (see Sect. 3.7).

If we introduce the Fourier transform $c(q)$ of $c(r)$ and use the convolution theorem we obtain

$$h(q) = \frac{c(q)}{1 - \rho_0 c(q)} \qquad (2.30)$$

which finally leads to

$$S(q) = \frac{1}{1 - \rho_0 c(q)}$$ (2.31)

2.5 Density Response Function

We can be interested in the density change due to the presence of an external potential

$$\mathcal{V}_{ext}(\mathbf{r}) = -\sum_\alpha \phi_{ext}(\mathbf{r} - \mathbf{r}_\alpha) .$$ (2.32)

The average density in the presence of $\mathcal{V}_{ext}(\mathbf{r})$ is given by[1]

$$\langle \rho(\mathbf{r}) \rangle_{\mathcal{V}_{ext}} = \frac{1}{Z_{\mathcal{V}_{ext}}} \int \frac{d^3\mathbf{r}'}{V} \prod_\alpha d^3\mathbf{r}_\alpha \underbrace{\sum_\alpha \delta(\mathbf{r} - \mathbf{r}_\alpha)}_{\rho(\mathbf{r})} e^{-\beta V\{\mathbf{r}_\alpha\}} e^{-\beta \mathcal{V}_{ext}(\mathbf{r}')}$$ (2.33)

with the configurational partition function as normalization factor

$$Z_{\mathcal{V}_{ext}} = \int \frac{d^3\mathbf{r}'}{V} \prod_\alpha d^3\mathbf{r}_\alpha e^{-\beta V\{\mathbf{r}_\alpha\}} e^{-\beta \mathcal{V}_{ext}(\mathbf{r}')}$$ (2.34)

and

$$V\{\mathbf{r}_\alpha\} = \frac{1}{2} \sum_{\alpha \neq \alpha'} \phi(|\mathbf{r}_\alpha - \mathbf{r}_{\alpha'}|)$$ (2.35)

Defining $\langle \rho \rangle_0$ to be the density average with $\mathcal{V}_{ext} = 0$ and going over to Fourier Transforms we obtain to lowest order in $\mathcal{V}_{ext} = 0$ (*linear response*)

$$\delta\rho(\mathbf{q}) = \langle \rho(\mathbf{q}) \rangle_{\mathcal{V}_{ext}} - \langle \rho \rangle_0 = \frac{\beta}{V} \langle \rho^*(\mathbf{q}) \rho(\mathbf{q}) \rangle \phi_{ext}(\mathbf{q}) \equiv \chi(q) \phi_{ext}(\mathbf{q})$$ (2.36)

the (static) *response function* or *susceptibility* is therefore given by

$$\chi(q) = \beta \rho_0 S(q)$$ (2.37)

This is a version of the famous *fluctuation-dissipation theorem*.

---

[1] $\beta = 1/k_B T$.

2.6   Mean Field Potential and Random Phase Approximation

In order to formulate an approximate theory for $S(q)$ it is useful to represent the interatomic interactions in terms of a *mean field potential* $U(q)$ which acts on the individual atoms as an *effective external polarization potential*

$$\phi_{\text{pol}}(\mathbf{q}) \equiv U(q)\delta\rho(\mathbf{q}). \qquad (2.38)$$

One then can use the *non-interacting Curie response function* $\chi_0 = -\beta\rho_0$ to write down the density change in terms of the real and the effective external potential:

$$\begin{aligned}
\delta\rho(q) &= \chi_0 \left[\phi_{\text{pol}}(\mathbf{q}) + \phi_{\text{ext}}(\mathbf{q})\right] \\
&= \chi_0 \left[U(q)\delta\rho(\mathbf{q}) + \phi_{\text{ext}}(\mathbf{q})\right] \\
&= \chi(\mathbf{q})\phi_{\text{ext}}(\mathbf{q}) \qquad (2.39)
\end{aligned}$$

from which follows

$$\chi(q) = \frac{\chi_0}{1 - \chi_0 U(q)} \qquad (2.40)$$

If we compare this with (2.31) we find

$$c(q) = -\beta U(q) \qquad (2.41)$$

We conclude that $-k_B T c(r)$ has the meaning of a *mean-field* potential. Identifying $U(r)$ with the true pairwise potential $\phi(r)$ is called the *Random-Phase approximation*. It gained its name from the theory of interacting electrons (or nucleons) [6]. There the RPA involves the decoupling of electronic correlation functions, which is only possible if the wave functions are assumed to have "random phases".

2.7   Integral Equation Theories for $g(r)$

We recall again the Ornstein–Zernike (OZ) relation between $g(r) = 1 + h(r)$ and the direct correlation function $c(r)$ (in a slightly modified form):

$$h(r) = c(r) + \rho_0 \int d^3\mathbf{r}' h(|\mathbf{r} - \mathbf{r}'|)c(r') \qquad (2.42)$$

The function $c(r)$ can, on the other hand, be calculated by functional integral and functional derivative techniques. Using such techniques and an appropriate diagram

formalism [3] one can come up with a second relation between $c(r)$ and $g(r)$ which is called the *closure relation* and constitutes a specific *integral equation theory* for $g(r)$. The most popular closure relations are
Percus–Yevick (PY):

$$c(r) = g(r)\left[1 - e^{\beta\phi(r)}\right] \tag{2.43}$$

Hypernetted-Chain (HNC):

$$c(r) = -\beta\phi(r) + h(r) - \ln g(r) \tag{2.44}$$

These closures together with the OZ relation constitute a self-consistent set of integral equations for $h(r)$ or $g(r)$.

2.8   PY Solution for Hard Spheres

We now consider the hard-sphere (HS) potential

$$\phi_{HS}(r) = \begin{cases} \infty & r < d \\ 0 & r > d , \end{cases} \tag{2.45}$$

where d is the *hard-sphere diameter*. In this case the PY integral equation can be solved *exactly*. The solution is given in terms of the *packing fraction*

$$\eta = \frac{\text{volume filled with spheres}}{\text{total volume}} = \frac{\pi}{6}d^3\rho_0 \tag{2.46}$$

and the dimensionless variable $x = r/d$

$$c(r) = \begin{cases} \lambda_1 - 6\eta\lambda_2 x + \frac{1}{2}\eta\lambda_1 x^3 & x < 1 \\ \\ 0 & x > 1 , \end{cases} \tag{2.47}$$

with

$$\lambda_1 = (1 + 2\eta)^2/(1 - \eta)^4 \tag{2.48a}$$

$$\lambda_2 = (1 + \frac{1}{2}\eta)^2/(1 - \eta)^4 \tag{2.48b}$$

If we compare the hard-sphere structure factors, plotted for different packing fractions η in Fig. 2.3 with the structure factors of liquid metals, the prominent examples of simple liquids (Figs. 2.2 and 2.5) we see that they are very similar. We comment on this in Sect. 2.11 (Fig. 2.5).

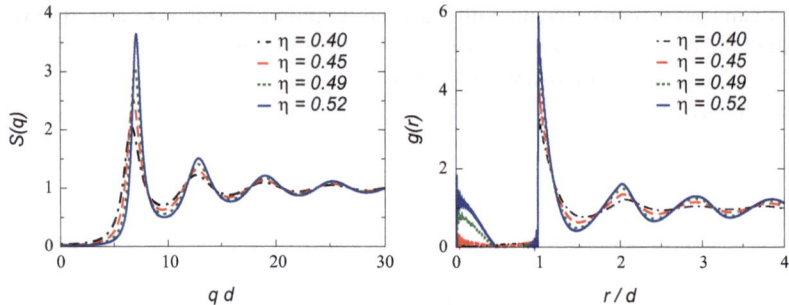

Fig. 2.3 Hard-sphere PY structure factors $S(q)$ and pair correlation functions $g(r)$

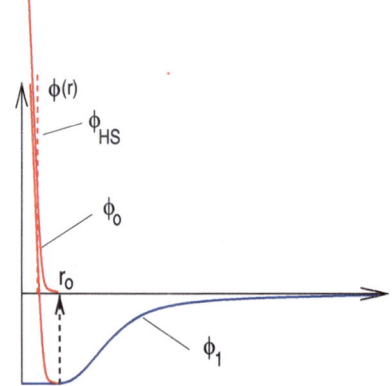

Fig. 2.4 Breakdown of a potential $\phi(r)$ into a hard-sphere-like part $\phi_0(r)$ and a longer-range part $\phi_1(r)$

2.9   Hard-Sphere Reference System

We now want to relate the structure of a simple liquid, which is described by a pairwise potential $\phi(r)$, in terms of the structure of hard spheres. For doing so we adopt the following procedure, due to Weeks et al. [9], Fig. 2.4. Denoting the separation at the minimum r_0 we define

$$\phi_0(r) = \begin{cases} \phi(r) - \phi(r_0) & r < r_0 \\ 0 & r > r_0 \, , \end{cases} \tag{2.49}$$

$$\phi_1(r) = \begin{cases} \phi(r_0) & r < r_0 \\ \phi(r) & r > r_0 \, . \end{cases} \tag{2.50}$$

We then relate a HS potential to ϕ_0. This can be done *"by hand"*, but a more systematical procedure is to require that some physical property of the hypothetical fluid described by $\phi_0(r)$ should coincide with a HS fluid with HS diameter d. One can demand that the spatial integral over $g_0(r)$ should coincide with that over $g_{HS}(r)$. This requirement is equivalent to demanding that the isothermal compressibilities of the two fluids should coincide.

2.10 Mean-Spherical Approximation

A quite successful integral equation theory for the structure of simple liquids is the *mean-spherical approximation* (MSA), which works with the procedure introduced in the last section and introduces the two constraints

$$g(r) = 0 \qquad\qquad r < d \qquad\qquad (2.51\text{a})$$

$$c(r) = -\frac{1}{k_B T}\phi_1(r) \qquad r > d \qquad\qquad (2.51\text{b})$$

i.e., we use a *random-phase approximation* for the *longer-range* part of the direct correlation function and use the short-range part for just demanding that $g(r)$ should be 0 for $r < d$. Together with the Ornstein–Zernike relation this constitutes an integral-equation theory for $g(r)$ which in many cases is even more successful as the PY or HNC theory. Of course it reduces to the PY theory for $\phi_1 = 0$.

2.11 Hard-Sphere Scaling of Liquid Metals

In Fig. 2.5 measured static structure factors of liquid metals [7] have been plotted against the wavenumber, multiplied by a suitable hard-sphere diameter d. The black line is the hard-sphere PY structure factor for the packing fraction $\eta = 0.45$. Obviously the structure is essentially determined by the hard core of the interatomic interactions. This hard core is produced by the Pauli exclusion principle, which forbids that the atomic orbitals of the metallic core electrons may overlap. It is further remarkable that almost all liquid metals near their melting point can be essentially described by a hard-sphere fluid with packing fraction $\eta = 0.45$ [1]. This is so, because there exists a maximum packing fraction for random close packing of hard spheres, which is smaller than the cubic or hexagonal crystalline close packing $\eta = 0.74$, namely $\eta = 0.634$ [8]. In this maximally packed state there is no mobility left, because the spheres can no more pass each other. This is the *glassy* or *jammed* state. In the liquid state of a material, which is dominated by short-ranged forces, like a liquid metal, the effective packing fraction must be smaller than this value. Obviously the equilibrium hard-sphere-like liquid needs a packing fraction as small as 0.45, and this is approximately the same for all liquid metals near the melting point.

Fig. 2.5 Measured structure factors of several liquid metals [7], together with the Perkus–Yevick hard-sphere structure factor (*black line*) for the packing fraction $\eta = 0.45$

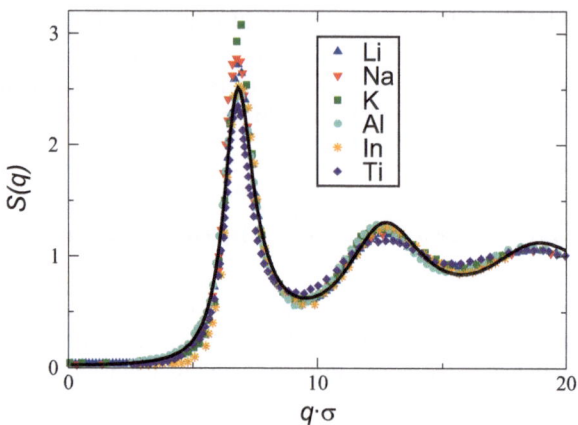

2.12   Perturbative RPA for the Compressibility of Liquids

As we have seen, the main features of the structure factors of simple liquids can be roughly accounted for by a HS structure. This means that the repulsive part of the potential $\phi_0(r)$ essentially acts as a HS potential, and that the long-ranged part $\phi_1(r)$ does not really play a major role for $S(q)$. However, this is only true for *finite* values of q. In the *long-wavelength* regime $q \ll q_0$, where q_0 marks the maximum of $S(q)$, the long-ranged part of the potential, ϕ_1 turns out to be important.

We have noted above, that in the long-wavelength limit $q \to 0$ we have

$$\lim_{q \to 0} S(q) = \rho_0 k_B T \kappa_T \tag{2.52}$$

So our statement involves the *hydrodynamic limit* of $S(q)$, in which this quantity is proportional to the compressibility. In this limit we can try the validity of the RPA—but not for the *entire* direct correlation function, but rather for the long-ranged part of the potential. So we write

$$c(r) = c_0(r) + c_1(r) \tag{2.53a}$$

$$c_0(r) = c_{HS}(r) \tag{2.53b}$$

with d chosen such that the ϕ_0 system has the same $S(q = 0)$ as the ϕ_{HS} system. For $c_1(r)$ we write

$$c_1(r) = -\frac{1}{k_B T} U_1(r) \tag{2.54}$$

For the mean-field potential $U_1(r)$ we may apply the *plain* RPA or the *extended* RPA (ERPA)

$$U_1(r) = \phi_1(r) \qquad \text{RPA}$$
$$U_1(r) = g(r)\phi_1(r) \qquad \text{ERPA} \qquad (2.55)$$

Both versions can be derived and motivated by a density functional formalism [2].

Inserting either of Eqs. (2.55) into (2.31) leads to

$$S(q = 0) = \frac{S_{HS}(q = 0)}{1 + S_{HS}(q = 0)\frac{\rho_0}{k_B T}U_1(q = 0)} \qquad (2.56)$$

with

$$U_1(q = 0) = \begin{cases} 4\pi \int_0^\infty dr\, r^2 \phi(r) & \text{RPA} \\ 4\pi \int_0^\infty dr\, r^2 g(r)\phi(r) & \text{ERPA} \end{cases} \qquad (2.57)$$

We noted previously that simple liquids—predominantly liquid metals—could be described near their melting point in terms of a HS fluid with packing fraction $\eta \approx 0.45$. In HS-PY approximation such a packing fraction leads to a value of $S(0)$ of ≈ 0.025. From (2.57) we expect that materials with a predominantly *attractive* long-range interaction $S(0)$ is *larger* than the HS value, and for the other case that it is *smaller*.

Let us take a closer look at the interatomic potentials $\phi(r)$ which characterize insulating liquids and metals.

The interatomic potentials $\phi(r)$ of *Insulating* simple liquids such as liquid rare gases or CH_4 are well described by the *Lennard–Jones* potential

$$\phi(r) = 4\epsilon \left[\left(\frac{\sigma}{r}\right)^{12} - \left(\frac{\sigma}{r}\right)^{6} \right]. \qquad (2.58)$$

If we subdivide this potential in the way described above into a hard-core (hard-sphere) part ϕ_0 and a long-range part ϕ_1 we realize that the latter is attractive in the whole r range, from which follows that both in the RPA ad ERPA the correction $U_1(0)$ is negative definite. This leads to an *increase* of $S(0)$ as compared to the hard-sphere value $S_{HS}(0)$ as can be verified from Table 2.1.

Let us now turn our attention to the interatomic potential of a metal which does not have unfilled d orbitals (i.e., *not* transition metals which have a complicated interatomic potential). In such *simple* metals the interatomic potential is a sum of a Coulomb repulsion of the ion cores of charge Ze (Z is the number of conduction electrons per atom or *valence*) and a contribution which describes the screening of the conduction electrons, featuring the electron-ion potential $v_i(q)$ and the electronic response function $\chi_e(q)$:

Table 2.1 The
long-wavelength limit of the
structure factor in insulating
and metallic liquids at their
triple points (taken from [2])

	$S_{RPA}(0)$	$S_{ERPA}(0)$	S_{exp}
Ne	0.056	0.055	0.052
Ar	0.044	0.044	0.050
Li	0.021	0.024	0.026
Na	0.024	0.024	0.023
K	0.026	0.027	0.024
Rb	0.027	0.028	0.022
Cs	0.028	0.028	0.024
Cu	0.021	0.021	0.021
Ag	0.022	0.022	0.019
Au	0.022	0.022	0.012
Mg	0.019	0.020	0.025
Zn	0.015	0.015	0.015
Cd	0.017	0.017	0.011
Hg	0.012	0.013	0.005
Al	0.014	0.014	0.017
Ga	0.007	0.007	0.005
In	0.012	0.012	0.006
Tl	0.015	0.015	0.010
Sn	0.010	0.009	0.007
Pb	0.012	0.012	0.009
Sb	0.010	0.010	0.020
Bi	0.008	0.008	0.009

$$\phi(r) = \frac{Z^2 e^2}{4\pi\epsilon_0 r} + \frac{1}{(2\pi)^3} \int d^3q \, e^{i\mathbf{q}\mathbf{r}} |v_i(q)|^2 \chi_e(q) \tag{2.59}$$

For most simple liquid metals at their melting points the electron gas has a Fermi distribution like that at $T = 0$, which has a sharp cutoff at the Fermi energy $E_F = \hbar^2 k_F/2m$, where $k_F = [3\pi^2 n]^{1/3}$ is the Fermi wavenumber, m is the electron mass and $n = Z\rho_0$ is the electron density. This is so, because the corresponding temperature (Fermi temperature) $T_F = E_F/k_B$ is of the order of 12,000 K. This leads to a sharp drop-off of $\chi_e(q)$ near $q = 2k_F$, which, in turn, leads in the r dependence of $\phi(r)$ to oscillations (*Friedel oscillations*) as depicted in Fig. 2.6. If $\phi(r)$ is now divided into the hard-core and long-range parts it turns out that the r integral of $r^2 \phi_1(r)$ takes a *positive* value, which is small for $Z = 1$ (alkali metals) but becomes large for $Z > 1$ (polyvalent metals). This leads to a *decrease* of the compressibility of the liquid metals as compared to the hard-sphere fluid, which can be checked in Table 2.1.

Fig. 2.6 Potentials $\phi(r)$ for
metals and insulators

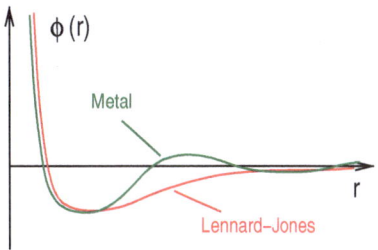

2.13 Relation to the van-der-Waals Equation of State

There exists a relation between the phenomenological equation of state of van der
Waals for the liquid-vapor phase separation and the perturbative RPA treated in the
previous section. The van-der-Waals equation of state for the pressure reads

$$\left(p + \frac{A}{V^2}\right)(V - B) = Nk_BT \tag{2.60a}$$

$$\text{or} \qquad p = -\frac{A}{V^2} + \frac{Nk_BT}{V - B} \tag{2.60b}$$

where A and B are phenomenological parameters called internal pressure and
covolume. This equation of state describes the quantities of state of expanded liquids
near their critical point very well, especially the liquid-vapor phase separation.

The perturbative RPA formula (2.57) can be written as

$$\frac{1}{S(0)} = -\frac{1}{k_BT}\frac{V^2}{N}\frac{\partial p}{\partial V} = \frac{1}{S_{HS}(0)} + \frac{U_1(0)}{k_BT} \tag{2.61}$$

We take the derivative of the van-der-Waals pressure with respect to the volume

$$\frac{\partial p}{\partial V} = \frac{2A}{V^3} - \frac{Nk_BT}{(V - B)^2} \tag{2.62}$$

and insert it into the definition of $1/S(0)$ displayed in the first part of (2.61):

$$\frac{1}{S(0)} = -\frac{1}{k_BT}\frac{V^2}{N}\frac{\partial p}{\partial V} = \frac{V^2}{(V - B)^2} - \frac{1}{k_BT}\frac{2A}{VN} \tag{2.63}$$

We see that this equation has a similar structure as the perturbative RPA: If we
identify the first term with a hard-core repulsion (as done, in fact, by van-der Waals),
the second one is the interaction term with interaction parameter $U(0) = -2A/VN$.
Therefore we expect that the perturbative RPA with negative (attractive) interaction
parameter $U(0)$ also describes a liquid-vapor phase separation.

It is in fact well known that a hard-sphere fluid does not show any liquid-vapor phase separation. Therefore it is called a "fluid". These considerations will become quite of importance in the next section which deals with binary solutions.

You might be asking how the liquid-vapor phase separation is to be described in the case of liquid metals, where, obviously, the interaction parameter is positive. The answer is that the equation of state of fluid metals near the critical point is much more complicated as that of insulating liquids. It does, in fact, *not* obey the law of corresponding states which can be described in terms of the van-der-Waals equation. Between the condensed liquid state and the vapor state there is a *metal-nonmetal transition*, which is until today not completely understood. The main feature of this transition is the breakdown of the screening of the conduction electrons due to the fact that at low densities the electron-electron repulsion becomes dominant and leads to a localization of the electrons near the ionic cores [5]. This leads to an instable regime in the pressure similar to the van-der-Waals mean-field model. We shall describe below for the case of metal-salt solutions, how the change of the metallic screening with concentration (metal density) can give rise to phase separation without a negative interaction parameter (Sect. 3.11).

2.14   The Resistivity of Liquid Metals

We wish to evaluate the resistivity of liquid metals in the Drude model of nearly free electrons. The resistivity is given in terms of the *relaxation time τ* or *relaxation rate* $1/\tau$ as follows

$$\rho = \frac{m}{ne^2} \frac{1}{\tau}. \tag{2.64}$$

τ is the mean time between collisions of electrons with ion cores, i.e., the relaxation rate is also the *collision* rate with the ion cores, which can be calculated from the transition probability $W_{\mathbf{k k_0}}$ from states $|\mathbf{k_0} >$ to states $|\mathbf{k} >$ as follows:

$$\frac{1}{\tau} = \frac{1}{(2\pi)^3} \int_{|\mathbf{k}|=k_F} d^3\mathbf{k} W_{\mathbf{k k_0}} (1 - \cos\theta) \tag{2.65}$$

where $\cos\theta$ is the angle between $\mathbf{k_0}$ and $\mathbf{k}$. Note that in the transport and scattering processes only electrons at the Fermi level are involved for which $|\mathbf{k}| = |\mathbf{k_0}| = k_F$ holds.

$W_{\mathbf{k k_0}}$ can be evaluated in Born approximation in a similar way as the neutron or X-ray scattering cross section. Denoting, again, $\mathbf{q} = \mathbf{k} - \mathbf{k_0}$ we have

$$W_{\mathbf{k k_0}} = \frac{2\pi}{\hbar} |< \mathbf{k}|\mathcal{V}_i|\mathbf{k_0} >|^2 = \frac{2\pi}{\hbar} n^2 S(q) |v_i(q)|^2 \tag{2.66}$$

Fig. 2.7 Temperature
dependence of $S(q)$ and the
positions of $2k_F$ for $Z = 1$
and $Z = 2$

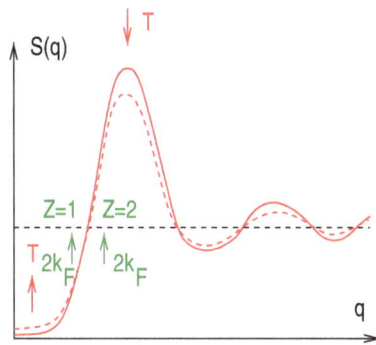

from which follows (taking in mind $|\mathbf{k}| = |\mathbf{k}_0| = k_F$)

$$\frac{1}{\tau} = \frac{1}{(2\pi)^3} \int_0^\pi d\theta \sin\theta W(q(\theta))[1 - \cos\theta] \qquad (2.67)$$

Going over to the integration variable

$$q(\theta) = \sqrt{k^2 + k_0^2 - 2kk_0 \cos\theta}$$

$$= \sqrt{2}k_F[1 - \cos\theta]^{1/2}$$

$$\Rightarrow \quad dq = \sqrt{2}k_F \frac{1}{[1 - \cos\theta]^{1/2}} \sin\theta d\theta$$

we have

$$\frac{1}{\tau} = \frac{1}{8\pi\hbar} \frac{n^2}{k_F^4} \int_0^{2k_F} dq q^3 S(q)|v_i(q)|^2 \qquad (2.68)$$

This formula is due to Edwards, Faber and Ziman [10–13]. The important
implication is that the *temperature dependence* is solely given by that of the static
structure factor. Moreover, the integral is strongly peaked at its maximum value
$q = 2k_F$, so we can expect that the temperature dependence of the resistivity of
liquid metals is essentially given by that of $S(q = 2k_F)$.

From Fig. 2.3 we see, how $S(q)$ depends on density, which decreases with
temperature. This temperature dependence is sketched also in Fig. 2.7. So for small
q $S(q)$ *increases*, for $q \approx q_0$ (peak position) it *decreases* with temperature. From
this it follows that for liquid *alkali* metals the resistivity increases with temperature
as it does usually also for solid metals. In solids this is due to the increased scattering
from phonons. For liquid *earth alkaline* metals and for *liquid alloys with* $Z = 2$
k_F turns out to approximately coincide with q_0. This is, in fact, observed in the
experimental data.

References

1. Ashcroft, N.W., Lekner, J.: Phys. Rev. **145**, 83 (1966)
2. Evans, R., Schirmacher, W.: J. Phys. C Solid State Phys. **11**, 2437 (1978)
3. Hansen, J.-P., McDonald, I.: Theory of Simple Liquids. Academic, New York (1986)
4. Lee, G.W., Gangopadhyay, A.K., Kelton, K.F., Hyers, R.W., Rathz, T.J., Rogers, J.R., Robinson, D.S.: Phys. Rev. Lett. **93**, 37802 (2004)
5. Mott, N.F., Davis, E.A.: Electronic Processes in Non-crystalline Materials. Clarendon, Oxford (1971)
6. Nozières, P., Pines, D.: Theory of Quantum Liquids. Benjamin, New York (1965)
7. Scopigno, T., Ruocco, G., Sette, F.: Rev. Mod. Phys. **77**, 881 (2005)
8. Song, C., Wang, P., Makse, H.A.: Nature **453**, 629 (2008)
9. Weeks, J.D., Chandler, D., Andersen, H.C.: J. Chem. Phys. **54**, 5237 (1971)
10. Edwards, S.: Philos. Mag. **3**, 1020 (1958)
11. Ziman, J. M.: Philos. Mag. **6**, 1013 (1961)
12. Ziman, J. M. and Faber, T. E.: Philos. Mag. **11**, 153 (1965)
13. Ziman, J. M.: Adv. Phys. **16**, 551 (1967)

Chapter 3
Structure and Thermodynamics of Binary Mixtures (Solutions)

3.1 General Definitions

As we now start dealing with mixtures of liquids we gradually cross the borderline between physics and chemistry. Therefore we must introduce the concept of *moles* which are lumps of $\mathcal{N}_{\mathrm{Avo}} = 6.022 \cdot 10^{23}$ particles (atoms or molecules). The number of moles is just $n = N/\mathcal{N}_{\mathrm{Avo}}$. We consider a mixture of two liquid phases A and B ("species") which are assumed to be in equilibrium with each other and consist of $N_A = n_A \mathcal{N}_{\mathrm{Avo}}$ and $N_B = n_B \mathcal{N}_{\mathrm{Avo}}$ particles, resp. If one is working at a given temperature T and pressure p (as we shall do) the appropriate thermodynamic potential is the Gibbs free enthalpy

$$G = H - TS = E - TS + pV . \tag{3.1}$$

where $H = E + pV$ is the enthalpy. The fundamental thermodynamic equation for G reads

$$dG = -S dT + V dp + \sum_{i=A,B} \mu_i dn_i \tag{3.2}$$

from which follows

$$S = -\frac{\partial G}{\partial T} \qquad \text{entropy} \tag{3.3a}$$

$$V = \frac{\partial G}{\partial p} \qquad \text{volume} \tag{3.3b}$$

$$\mu_i = \frac{\partial G}{\partial n_i} \qquad \text{chemical potentials} \tag{3.3c}$$

W. Schirmacher, *Theory of Liquids and Other Disordered Media*, Lecture Notes in Physics 887, DOI 10.1007/978-3-319-06950-0_3,
© Springer International Publishing Switzerland 2015

For any physical variable X (like G, E, T, S, p, and V) that characterizes the total mixture one can introduce so-called *partial* quantities which are defined as

$$x_i = \frac{\partial X}{\partial n_i} \qquad i = A, B \qquad (3.4)$$

from which follows (this is an exercise to be done!)

$$X = n_A x_A + n_B x_B \qquad (3.5)$$

X is also called an *extensive* quantity, and the x_i are the corresponding *intensive* quantities. We identify immediately the chemical potentials μ_i as the partial free enthalpies of the mixture. We now introduce the so-called *concentrations* or *mole fractions* $c_i = N_i/N = n_i/n$. Since $c_A + c_B = 1$ only one of these variables is independent, and we define $c_A \equiv c$ to be the independent variable, so that $c_B = 1 - c$. For any extensive quantity X we have (exercises!)

$$\frac{\partial X}{\partial c} = x_A - x_B \qquad (3.6)$$

and

$$x_A \frac{\partial x_A}{\partial c} + x_B \frac{\partial x_B}{\partial c} = 0. \qquad (3.7)$$

In particular, we have

$$x_A \frac{\partial \mu_A}{\partial c} + x_B \frac{\partial \mu_B}{\partial c} = 0, \qquad (3.8)$$

which is the famous *Gibbs–Duhem* relation.

3.2   Cross-Sections and Partial Correlation Functions

Let us recall the scattering cross-section for energy unresolved neutron or X-ray scattering from a liquid

$$\frac{d\sigma}{d\Omega} = \left\langle \sum_{\alpha,\beta=1}^{N} f_\alpha^*(q) f_\beta(q) e^{i\mathbf{q}\cdot[\mathbf{r}_\alpha - \mathbf{r}_\beta]} \right\rangle, \qquad (3.9)$$

where we now have labelled the form factors with atomic indices α, β. Let us assume, that the atoms now belong to either species A or species B. We would like to factorize (3.9) into a form and structure factor as done with the expression (14) for the mono-atomic liquid. However, due to the heterogeneous character of a mixture such a factorization is no more possible. All we can do is to come up with a *linear combination* of form and structure factors. In order to derive such an expression we again separate the terms into self and distinct parts. Let us keep in mind that the distinct correlation function $h(q)$ was defined as

$$h(q) = \frac{1}{\rho_0}[S(q) - 1] = \frac{V}{N^2}\left\langle \sum_{\alpha \neq \beta} e^{i\mathbf{q}\cdot[\mathbf{r}_\alpha - \mathbf{r}_\beta]} \right\rangle. \tag{3.10}$$

We now define the corresponding quantities in which the α, β sums are only over A- or B-type atoms:

$$h_{ij}(q) = \frac{V}{N_i N_j}\left\langle \underbrace{\sum_{\alpha=1}^{N_i} \sum_{\beta=1}^{N_j}}_{\alpha \neq \beta} e^{i\mathbf{q}\cdot[\mathbf{r}_\alpha^{(i)} - \mathbf{r}_\beta^{(j)}]} \right\rangle \qquad i, j = \text{A or B} \tag{3.11}$$

and obtain, since the incoherent $\alpha = \beta$ terms occur only linearly in the concentrations c_i

$$\frac{d\sigma}{d\Omega} = N\left(c_A|f_A(q)|^2 + c_B|f_B(q)|^2 + \rho_0 \sum_{i=A,B} \sum_{j=A,B} c_i c_j f_i^*(q) f_j(q) h_{ij}(q) \right) \tag{3.12}$$

In the same way the Edwards–Faber–Ziman formula for a liquid metallic alloy is given by

$$\rho = \frac{m}{ne^2}\frac{1}{8\pi\hbar}\frac{n^2}{k_F^4}\int_0^{2k_F} dq\, q^3 \left(c_A|v_A(q)|^2 + c_B|v_B(q)|^2 \right.$$

$$\left. + \rho_0 \sum_{i=A,B} \sum_{j=A,B} c_i c_j v_i^*(q) v_j(q) h_{ij}(q) \right) \tag{3.13}$$

One defines the so-called *partial structure factors* as

$$S_{ij}(q) = 1 + \rho_0 h_{ij}(q) \qquad i, j = \text{A or B}, \tag{3.14}$$

but, as we see from (3.12) and (3.13), this definition is not of much use, as the functions $h_{ij}(q)$ and not $S_{ij}(q)$ enter into the expressions. On the other hand, the Fourier transforms of $h_{ij}(q)$ are related to the *partial radial distribution functions* $g_{ij}(r)$

$$h_{ij}(r) = \frac{1}{2\pi^2 r}\int_0^\infty dq\, q\, \sin(q)h(q) = g_{ij}(r) - 1 \tag{3.15}$$

$\rho_0 4\pi r^2 g_{ij}(r)dr$ gives the probability for the presence of a j particle inside a spherical shell of thickness dr, if there is an i particle at the origin.

3.3 Number and Concentration Fluctuations

Instead of working with the partial structure factors $S_{ij}(q)$ or the functions $h_{ij}(q)$ one can define linear combinations of these functions which are the correlation functions of the density fluctuations $\delta\rho$ and the concentration fluctuations δc [1,2]:

$$S_{\rho\rho}(q) = c_A^2 S_{AA}(q) + c_B^2 S_{BB}(q) + 2c_A c_B S_{AB}(q) \tag{3.16a}$$

$$S_{\rho c}(q) = c_A c_B \{c_A [S_{AA}(q) - S_{AB}(q)] - c_B [S_{BB}(q) - S_{AB}(q)]\} \tag{3.16b}$$

$$S_{cc}(q) = c_A c_B \{1 + c_A c_B [S_{AA}(q) + S_{BB}(q) - 2S_{AB}(q)]\} \tag{3.16c}$$

In terms of these quantities (3.12) and (3.13) take the form

$$\frac{d\sigma}{d\Omega} = N \left(\left|\overline{f}\right|^2 S_{\rho\rho}(q) + |f_A - f_B|^2 S_{cc}(q) + 2\overline{f}^*(f_A - f_B)S_{\rho c}(q) \right) \tag{3.17}$$

with $\overline{X} \equiv c_A X_A + c_B X_B$.

$$\rho = \frac{m}{ne^2} \frac{1}{8\pi\hbar} \frac{n^2}{k_F^4} \int_0^{2k_F} dq\, q^3 \Big[|\overline{v}|^2 S_{\rho\rho}(q) + |v_A - v_B|^2 S_{cc}(q)$$

$$+ 2Re\{\overline{v}^*(v_A - v_B)\}S_{\rho c}(q) \Big] \tag{3.18}$$

At $q = 0$ the following relations hold:

$$S_{\rho\rho}(0) = \frac{1}{N}\langle(\Delta N)^2\rangle = \theta + \delta^2 S_{cc}(0) \tag{3.19a}$$

$$S_{\rho c}(0) = \langle\Delta N \Delta c\rangle \quad = -\delta S_{cc}(0) \tag{3.19b}$$

$$S_{cc}(0) = N\langle(\Delta c)^2\rangle = k_B T / g_{cc} \tag{3.19c}$$

with the three thermodynamic quantities

$$\theta = \rho_0 k_B T \kappa_T \tag{3.20a}$$

$$\delta = \frac{1}{V}\left(\frac{\partial V}{\partial c}\right)_{P,T,N} = \frac{v_A - v_B}{n_A v_A + n_B v_B} \tag{3.20b}$$

$$g_{cc} = \frac{1}{N}\left(\frac{\partial^2 G}{\partial c^2}\right)_{P,T,N} \tag{3.20c}$$

Here θ is again related to the isothermal compressibility κ_T, v_i are the partial molar volumina and g_{cc} is the *stability function*. The *free enthalpy of mixing* is defined to be

$$\Delta G = G - N_A G_A^{(0)} - N_B G_B^{(0)} \tag{3.21}$$

Fig. 3.1 *Red* and *blue*
particles in the demixed and
mixed state

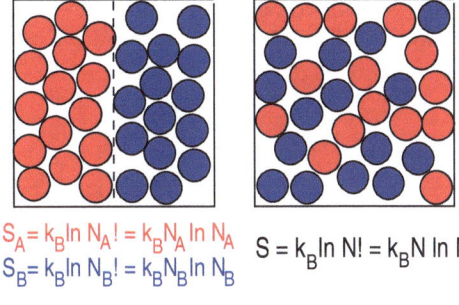

$S_A = k_B \ln N_A! = k_B N_A \ln N_A$
$S_B = k_B \ln N_B! = k_B N_B \ln N_B$ $S = k_B \ln N! = k_B N \ln N$

Fig. 3.2 Entropy as a
function of concentration

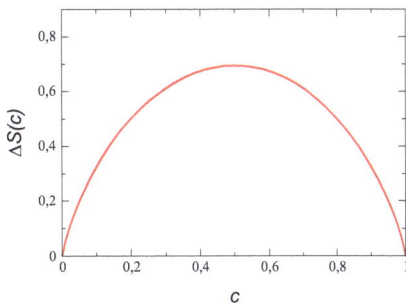

where $G_i^{(0)}$ are the free enthalpies of the pure systems. As these quantities do not
depend on concentration we only need to know ΔG for calculating g_{cc}.

3.4 Entropy of Mixing

Let us perform the classical gedanken experiment (Fig. 3.1) for defining the entropy
and showing that the entropy increases with mixing. For a system of N particles
the entropy is defined to be k_B times the logarithm of the number of possible
configurations. We imagine that the particles can be distributed into N volume
"cells" in $N!$ different ways. Then the entropy is just

$$S = k_B \ln N! \overset{N \to \infty}{=} k_B N \ln N \qquad (3.22)$$

We now start the gedanken experiment at the left side of Fig. 3.1. The Entropy is
initially the sum of the red and blue terms. If we achieve a complete random mixture
the entropy is given by the black expression. The increase in entropy is

$$\Delta S^{(0)} = k_B [(N_A + N_B) \ln N - N_A \ln N_A - N_B \ln N_B]$$
$$= -k_B N [c_A \ln c_A + c_B \ln c_B] \qquad (3.23)$$

The Graph of $\Delta S^{(0)}(c)$ is shown in Fig. 3.2.

For non-interacting particles (for which $\Delta H = 0$) the stability function obviously is given by

$$g_{cc} = \frac{1}{N}\frac{\partial^2}{\partial c^2}[-T\Delta S] = k_B T \frac{\partial^2}{\partial c^2}[c\ln c + (1-c)\ln(1-c)]$$

$$= k_B T\left[\frac{1}{c} + \frac{1}{(1-c)}\right] = k_B T\left[\frac{1}{c(1-c)}\right] \quad (3.24a)$$

$$\Rightarrow \qquad S_{cc}(0) = c(1-c) \qquad (3.24b)$$

3.5   Partial Structure Factors of Ideal Solutions

Ideal solutions are defined to be systems in which there are either no interactions or all interactions are equal so that there are no *excess* interactions:

$$\phi_{AA}(r) = \phi_{BB}(r) = \phi_{AB}(r) \qquad (3.25)$$

In such a system the partial structure factors S_{ij} are all equal and we have

$$S^{(0)}_{\rho\rho}(q) \equiv S(q) \qquad S^{(0)}_{\rho c}(q) = 0 \qquad S^{(0)}_{cc}(q) = c(1-c) \qquad (3.26)$$

Inserting this into (3.17) we obtain

$$\frac{d\sigma}{d\Omega} = \left|\overline{f}\right|^2 S(q) + |f_A - f_B|^2 c(1-c) \qquad (3.27)$$

One obtains the same expression from (3.12) if all h_{ij} are set equal to $\frac{1}{\rho_0}[S(q) - 1]$. In the case of X-ray diffraction on weakly interacting alloys the second term is just a *background* which is q independent and is called *Laue background*. Such a background is also observed in randomly mixed crystals.

3.6   Direct Correlation Functions

As in the single-component case one can define *direct correlation functions* by the equations

$$h_{ij}(r) = c_{ij}(r) + \rho_0 \sum_{\ell=A,B} c_\ell \int d^3r' h_{i\ell}(r')c_{\ell j}(|\mathbf{r} - \mathbf{r}'|) \qquad (3.28)$$

By applying the convolution theorem and a 2×2 matrix inversion one can derive the following relations between their Fourier transforms $C_{ij}(q) \equiv \rho_0 d^3 \mathbf{r} e^{i \mathbf{q} \mathbf{r}} c_{ij}(r)$ and the number and concentration structure factors:

$$S_{\rho\rho}(q) = \Theta(q) + \Delta^2(q) S_{cc}(q) \tag{3.29a}$$

$$S_{\rho c}(q) = -\Delta(q) S_{cc}(q) \tag{3.29b}$$

$$S_{cc}(q) = \left[\frac{1}{c_A c_B} - \frac{\Delta(q)^2}{\Theta(q)} - C_{AA}(q) - C_{BB}(q) + 2C_{AB}(q) \right]^{-1} \tag{3.29c}$$

$$\Theta(q) = \left[1 - c_A^2 C_{AA}(q) - c_B^2 C_{BB}(q) - 2c_A c_B C_{AB}(q) \right]^{-1} \tag{3.29d}$$

$$\Delta(q) = \Theta(q) \left[c_A \left(C_{AA}(q) - C_{AB}(q) \right) - c_B \left(C_{BB}(q) - C_{AB}(q) \right) \right] \tag{3.29e}$$

Here the quantities $\Theta(q)$ and $\Delta(q)$ are the generalizations of $\theta = \Theta(q = 0)$ and $\delta = \Delta(q = 0)$ of (3.20).

3.7   Perturbative RPA for $q = 0$ and Regular Solution Model

As in the single-component case we now generalize the perturbative RPA as follows

$$C_{ij}(q = 0) = C_{ij}^{(0)} - \frac{1}{k_B T} U_{ij}^{(1)} \tag{3.30}$$

with

$$U_{ij}^{(1)} = \begin{cases} \rho \int d^3 \mathbf{r} \phi_{ij}(r) & RPA \\ \\ \rho \int d^3 \mathbf{r} g_{ij}(r) \phi_{ij}(r) & ERPA \end{cases} \tag{3.31}$$

We now define a *Flory–Huggins interaction parameter* [5,8] by

$$\omega = k_B T \chi = U_{AB}^{(1)} - \frac{1}{2} \left[U_{AA}^{(1)} + U_{BB}^{(1)} \right] \tag{3.32}$$

and obtain for $S_{cc}(0)$

$$S_{cc}(0) = \left[\frac{1}{S_{cc}^{(0)}} - \frac{\delta_0^2}{\theta_0} + \frac{\delta_1^2}{\theta_1} - \frac{1}{k_B T} 2\omega \right]^{-1}, \tag{3.33}$$

Fig. 3.3 $S_{\rho\rho}(q)$ and the different possible forms of $S_{cc}(q)/c_A c_B$ for different types of chemical order

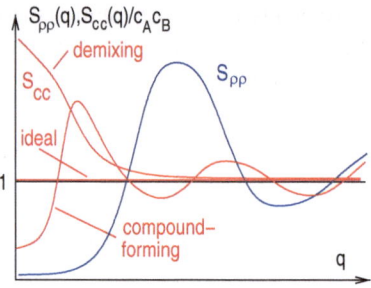

where the index 0 refers to the hard-core system, and the index 1 refers to the combination of the direct correlation functions $C_{ij}^{(1)}(q = 0) \equiv -U_{ij}^{(1)}(q = 0)/k_B T$ given by (3.29d) and (3.29e).

In the case $\delta = 0$, $S_{cc}^{(0)}(0) = c_A c_B$ we have

$$
S_{cc}(0) = \left[\frac{1}{c_A c_B} - \frac{1}{k_B T} 2\omega \right]^{-1}
\tag{3.34}
$$

The quantity $S_{cc}(0)$ can be taken as a parameter which describes the *degree of chemical ordering* (Fig. 3.3). In the non-interacting case, which is equivalent to the case in which all three pair potentials are equal, there is no chemical ordering and we have $S_{cc}(0) = c_A c_B$. If the $A - B$ potential is more *attractive* than the average of the $A - A$ and $B - B$ potentials we have a tendency towards forming an $A - B$ compound, and there will be a *chemical ordering* as in an antiferromagnet. This establishes a *superstructure* (Überstruktur) and is accompanied by a *pre-peak* in $S_{cc}(q)$, which is situated half-way between the principle peak of $S_{\rho\rho}(q)$ and $q = 0$. On the other hand, if the $A - B$ potential is *less attractive* than the average of the $A - A$ and $B - B$ potentials we have a tendency towards de-mixing, which can result in a demixing phase transition if the quantity $\chi = \omega/k_B T$ becomes too large.

A tendency towards demixing is accompanied by an enhanced *small-angle scattering*, i.e., an increase in the low-q part of $S_{cc}(q)$.

We now call a model substance in which $\delta = 0$, $S_{cc}^{(0)} = c_A c_B$ holds, a *regular solution*. Inserting (3.33) into (3.19c) and (3.20c) we obtain (with $c \equiv c_A$) for the stability function

$$
g_{cc} = \frac{1}{N} \left(\frac{\partial^2 \Delta G}{\partial c^2} \right)_{P,T,N} = -2\omega + k_B T \frac{1}{c(1-c)}
\tag{3.35}
$$

If we integrate (3.35) twice with the boundary condition $\Delta G(c = 0) = \Delta G$ $(c = 1) = 0$ we obtain the *free enthalpy of mixing for a regular solution*

$$
\frac{1}{N} \Delta G = c(1-c)\omega + k_B T \left[c \ln c + (1 - c) \ln(1 - c) \right]
\tag{3.36}
$$

3.8 Activities and Activity Coefficients

The total Gibbs free enthalpy of an *ideal* solution, i.e., a regular solution with $\omega = 0$, can be written as

$$G = n_A \mu_A + n_B \mu_B = cG(c = 0) + (1 - c)G(c = 1)$$
$$+ nRT \left[c \ln c + (1 - c) \ln(1 - c) \right] , \qquad (3.37)$$

were $R = \mathcal{N}_{\text{Avo}} k_B$ is the gas constant. Therefore the chemical potentials have the form

$$\mu_i = \mu_i^{(0)} + RT \ln c_i , \qquad (3.38)$$

where $\mu_i^{(0)} = G(c_i = 1)/n$, n being the number of moles. In a *regular* solution one defines the *absolute activities* a_A and a_B as

$$\mu_i = \mu_i^{(0)} + RT \ln a_i \qquad (3.39)$$

so that we have

$$\frac{1}{S_{cc}(q = 0)} \frac{\partial^2}{\partial c^2} \left(\frac{\Delta G}{nRT} \right) = \frac{1}{RT} \frac{\partial}{\partial c}(\mu_A - \mu_B) = \frac{1}{RT} \frac{1}{1 - c} \frac{\partial}{\partial c} \mu_A = \frac{1}{1 - c} \frac{\partial \ln a_A}{\partial c} \qquad (3.40)$$

The third equality follows from the Gibbs-Duhem equality $\frac{\partial}{\partial c} \mu_B = -\frac{c}{1-c} \frac{\partial}{\partial c} \mu_A$. These coefficients "mimic" the thermodynamic laws for ideal mixtures, in which one has to replace the c_i by the a_i.

One can, furthermore, consider the ratio of the absolute activities, and their ideal values, the concentrations

$$f_i = a_i / c_i \qquad (3.41)$$

These are the *activity coefficients*. the free enthalpy of mixing becomes

$$\frac{\Delta G}{nRT} = [c \ln c + (1 - c) \ln(1 - c)] + [c \ln f_A + (1 - c) \ln f_B] \qquad (3.42)$$

So we can write for the *enthalpy of mixing*

$$\frac{\Delta H}{nRT} = c(1 - c)\chi = [c \ln f_A + (1 - c) \ln f_B] \qquad (3.43)$$

We identify the quantities $RT \ln f_i$ as the *partial enthalpies*. If the Flory–Huggins-Parameter χ does not depend on concentration we have

$$\frac{\partial^2}{\partial c^2}\left(\frac{\Delta H}{nRT}\right) = -2\chi = \frac{1}{1-c}\frac{\partial \ln f_A}{\partial c} \tag{3.44}$$

3.9 Partial Vapor Pressures Above a Regular Solution

Let us suppose we have one mole of a binary liquid mixture inside a closed container. In equilibrium the chemical potentials of the vapor and that of the liquid *of both species* should be the same. Let us attach a semi-permeable membrane to the vessel and change the partial pressure of species i. The change in the chemical potential of the vapor of i, i.e., the change in the partial Gibbs free enthalpy will be

$$\Delta \mu_i = \int_{p_{\text{ref}}}^{p_i} V(p)\mathrm{d}p . \tag{3.45}$$

Inserting the ideal gas law $V(p) = RT/p$ we obtain

$$\Delta \mu_i = RT \ln p_i/p_{\text{ref}} \tag{3.46}$$

so that we can write

$$\mu_i = \mu_{i,\text{ref}} + RT \ln p_i/p_{\text{ref}} . \tag{3.47}$$

Taking for the reference chemical that of the pure species $\mu_i^{(0)}$ and comparing with expression (3.38) for the chemical potential of an ideal solution we obtain *Raoult's law* for the partial vapor pressure of an ideal solution

$$p_i = c_i\, p_0 \tag{3.48}$$

For a *regular* solution we obtain by comparing with (3.39)

$$p_i = a_i\, p_0 \tag{3.49}$$

3.10 Phase Separation in Regular Solutions

We now consider a regular solution with a free enthalpy of mixing (Fig. 3.4)

$$\frac{1}{N}\Delta G = c(1-c)\omega + k_B T\left[c \ln c + (1-c)\ln(1-c)\right] \tag{3.50}$$

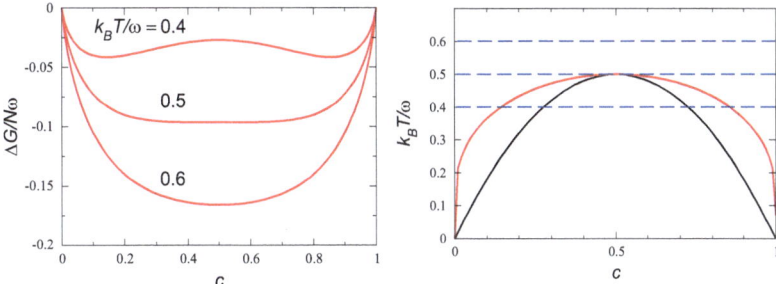

Fig. 3.4 *Left*: free enthalpy of mixing for the three temperatures $k_B T/\omega = 0.4, 0.5(T_C)$, and 0.6. *Right*: the phase diagram with *spinodal lines* and equilibrium concentrations. The *dashed lines* give the temperatures of the left figure

and a corresponding stability function

$$g_{cc} = \frac{1}{N}\left(\frac{\partial^2 \Delta G}{\partial c^2}\right)_{P,T,N} = -2\omega + k_B T\frac{1}{c(1-c)} \tag{3.51}$$

We immediately notice that there occurs an instability, once $k_B T/2\omega \le c(1-c)$. Because $c(1-c)$ does not exceed 0.25, the critical temperature for separation into a B-rich and an A-rich phase is given by $k_B T_c/\omega = 0.5$. Below this temperature the line of instability, given by

$$\frac{k_B T}{\omega} = \frac{1}{\chi} = 2c(1-c) \tag{3.52}$$

is called *spinodal line*. The equilibrium concentrations of the *A*-rich and *B*-rich phases are obtained by the condition that the chemical potentials of both species must coincide:

$$\mu_{A,\text{rich}} = \mu_{A,\text{poor}} \qquad \mu_{B,\text{rich}} = \mu_{B,\text{poor}} \tag{3.53}$$

If these equations hold, we must also have

$$\mu_{A,\text{rich}} - \mu_{B,\text{rich}} = \mu_{A,\text{poor}} - \mu_{B,\text{poor}} \tag{3.54}$$

Because the chemical potential difference is just proportional to the *slope* of the curve $\Delta G(c)$, the two equilibrium concentrations are given by the *double-tangent construction*: in the instable region, where $\Delta G(c)$ varies non-monotonically with concentration one seeks a line which touches the curve at two points (double tangent). In our case these concentrations co-incide with the positions of the minima of $\Delta G(c)$ and are given by

$$k_B T/\omega = \frac{2c - 1}{\ln c - \ln(1-c)} \tag{3.55}$$

3.11   Phase Separation in Metal-Salt Solutions

We consider a mixture of a metal with a halogen which we model by the following set of interactions

$$\phi_{ij}(r) = \phi_{ij}^{(0)}(r) + Q_i Q_j e^2 \frac{e^{-\lambda r}}{r} \qquad i = A, B \qquad (3.56)$$

$\phi_{ij}(0)(r)$ is a hard-core potential and the rest is a screened Coulomb potential between charges $Q_i e$ which are supposed to obey the global neutrality condition

$$c_A Q_A + c_B Q_B = 0 \qquad (3.57)$$

Although such a material is no more a system which is "weakly interacting" we can still try to use the RPA, because the screened Coulomb terms are much smaller than the repulsive ones. The interaction energy is

$$\omega(c) = -\frac{1}{2} e^2 \frac{c_B}{c_A} Q_B^2 \rho_0 \int_{r>d} d^3 \mathbf{r} \frac{1}{r} e^{-\lambda r} \qquad (3.58)$$

If we transform now from the concentrations c_A, c_B to the variable x which appears in the chemical formula for the "pseudo-binary" metal salt mixture $M_x(MH)_{1-x}$ via $c_A x = 1 - 2c_B$ we arrive at an equation for the interaction energy [6, 7]

$$\omega(x) = -A(1 - x)e^{\lambda(x)d} \qquad (3.59)$$

Here A is an interaction parameter, which can in principle be calculated, using the RPS or ERPA, and we have assumed that the main contribution to the interaction comes from the contact value $r = d$.

We now introduce the essential feature of this system, namely the *concentration dependence of the screening* λ by assuming it to vary linearly around some arbitrary intermediate concentration x_0

$$\lambda(x) = \lambda(x_0) + \frac{1}{d}\lambda'(x - x_o) \qquad (3.60)$$

where λ'/d is the linear Taylor coefficient. The enthalpy of mixing of the pseudobinary system with this $\lambda(x)$ is

$$\Delta H(x) = E(x) - (1 - x)E(0) - xE(1) = A(1 - x)\left[1 - e^{-\lambda' x}\right] \qquad (3.61)$$

We now take for the entropy that of the ideal pseudobinary system and define reduced quantities

$$f = [\Delta H - \Delta S(x)T]/A \equiv (1 - x)\left[1 - e^{-\lambda' x}\right] - ts(x) \qquad (3.62)$$

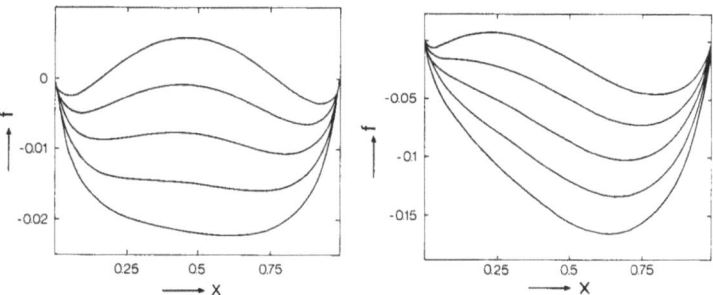

Fig. 3.5 Free enthalpy isotherms $f(x)$ for $\lambda' = 0.2$ (*left*) and $\lambda' = 2.0$ (*right*) from Holzhey and Schirmacher [6]

with $t = T/A$ and

$$s(x) = -x \ln x - (1 - x) \ln(1 - x) \tag{3.63}$$

We have calculated the free enthalpy isotherms $f(x)$ and phase diagrams for a *weak* density dependence of the screening ($\lambda' = 0.2$) and a *strong* one ($\lambda' = 2.0$) (Fig. 3.5). This may be thought to come from different electronegativities between the metal and the halide. It should be emphasized that in this model the interaction energy is *attractive* and the instability in the free enthalpy results entirely from the concentration dependence of the screening length $\lambda(x)$. We see that the strong density dependence of the screening results in an *asymmetric* phase diagram, a week density dependence in a symmetric one. In the measured phase diagrams (Fig. 3.6) we see that the fluorine-potassium system has a very asymmetric one, whereas the iodine-potassium diagram is symmetric. It is, in fact, known that the electronegativity difference between I and K is much less than that between F and K.

3.12 Integral Equation Theories for $g_{ij}(r)$

Until now we have only exploited the thermodynamic $q \rightarrow o$ limit of the partial structure factors $S_{ij}(q) = \rho_0 h_{ij}(q)$ and the related quantities $S_{\rho\rho}(q), S_{\rho c}(q), S_{cc}(q)$. In the same way as done for the single-component case we can set up closure relations for the partial direct correlation functions, defined by

$$g_{ij}(r) - 1 = h_{ij}(r) = c_{ij}(r) + \rho_0 \sum_{\ell=A,B} c_\ell \int d^3r' h_{i\ell}(r') c_{\ell j}(|\mathbf{r} - \mathbf{r}'|) \tag{3.64}$$

Percus–Yevick (PY):

$$c_{ij}(r) = g_{ij}(r) \left[1 - e^{\beta \phi_{ij}(r)}\right] \tag{3.65}$$

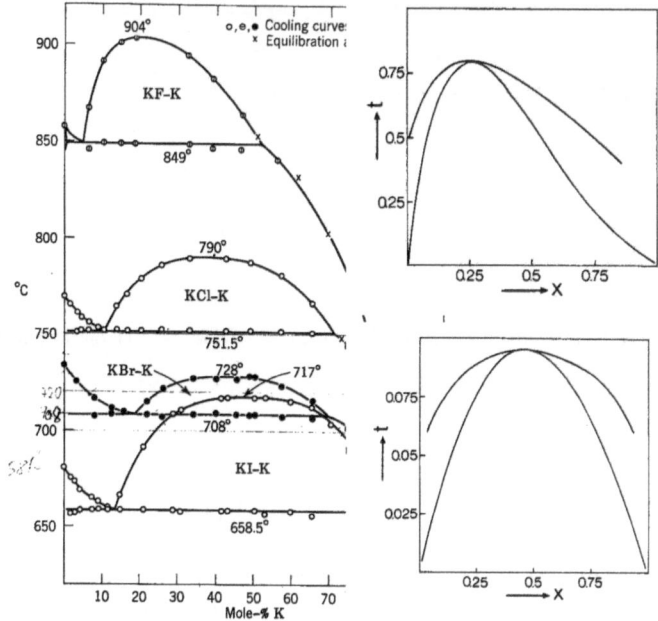

Fig. 3.6 Phase diagrams for the sodium halides [3] together with the model phase diagrams, for $\lambda' = 2.0$ (*top*) and $\lambda' = 0.2$ from [7]

Hypernetted-Chain (HNC):

$$c_{ij}(r) = -\beta\phi_{ij}(r) +_{ij} h(r) - \ln g_{ij}(r) \tag{3.66}$$

Mean-Spherical Approximation (MSA):

$$g_{ij}(r) = 0 \qquad\qquad r < d \tag{3.67a}$$

$$c_{ij}(r) = -\frac{1}{k_B T}\phi_{ij}^{(1)}(r) \qquad r > d \tag{3.67b}$$

The PY approximation can—as in the single-component case—be solved analytically for a hard-sphere potential with hard-sphere diameters d_{ij} [9]. If these do not differ too much from each other, the partial structure factors are similar to each other, too, so that the quantities $S_{pc}(q)$ and $S_{cc}(q) - c_A c_B$ are quite small in the whole q range.

As indicated in the beginning, the physics (and physical chemistry) of binary mixtures is dominated by the *ordering potential*

$$\phi_{\text{ord}}(r) = -\frac{1}{2}\phi_{cc}(r) = \phi_{AB}(r) - \frac{1}{2}[\phi_{AA}(r) + \phi_{AA}(r)] \tag{3.68}$$

The finite-q version of the regular solution model is a material with

$$\Delta(q) \propto c_A \left(C_{AA}(q) - C_{AB}(q) \right) - c_B \left(C_{BB}(q) - C_{AB}(q) \right) = 0 \tag{3.69}$$

and

$$S_{\rho c}(q) \propto c_A \left[S_{AA}(q) - S_{AB}(q) \right] - c_B \left[S_{BB}(q) - S_{AB}(q) \right] = 0. \tag{3.70}$$

In this case the $S_{\rho\rho}(q)$ and $S_{cc}(q)$ decouple completely. One can consider the case in which the short-range potentials $\phi_{ij}^{(0)}$ are equal and for the long-range potentials only the $\rho\rho$ and the cc combinations (3.68) are non-zero. In this case the function $S_{\rho\rho}(q)$ is completely fixed by the short-range potential, which can be represented by that of a single-component model, i.e., the hard-sphere Percus–Yevick model.

For the function

$$h_{cc}(r) = \frac{\rho_0}{(2\pi)^3} \int d^3 \mathbf{r}\, e^{-i\mathbf{q}\mathbf{r}} \left[\frac{S_{cc}(q)}{c_A c_B} - 1 \right] \tag{3.71}$$

the MSA takes the form

$$h_{cc}(r) = 0 \qquad\qquad r < d \tag{3.72a}$$

$$c_{cc}(r) = \frac{2}{k_B T} \phi_{\text{ord}}^{(1)}(r) \qquad r > d \tag{3.72b}$$

with

$$c_{cc}(r) = c_{AA}(r) + c_{BB}(r) - 2c_{AB}(r), \tag{3.73}$$

and we have for the Fourier transform

$$\frac{c_A c_B}{S_{cc}(q)} = 1 - c_A c_B \rho_0 c_{cc}(q) = 1 - c_A c_B \left[C_{AA}(q) + C_{BB}(q) - 2C_{AB}(q) \right] \tag{3.74}$$

As in the single-component case this version can be solved for a screened coulomb potential

$$\phi_{\text{ord}}^{(1)}(r) \propto \frac{1}{r} e^{-\lambda r}. \tag{3.75}$$

3.12.1   The Liquid Alloy Li₄Pb

In the case of the liquid alloy Li$_4$Pb one can prepare an isotopic mixture in such a way that the measured diffraction intensity is just proportional to $S_{cc}(q)$ because

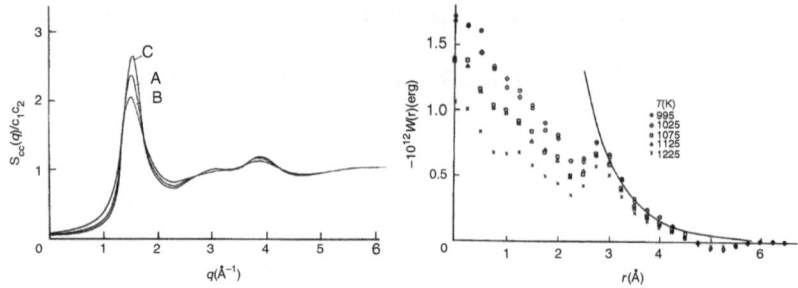

Fig. 3.7 *Left*: $S_{cc}(q)$ for A T $=$ 1,075 K, B T $=$ 1,225 K, C T $=$ 995 K, [10]. *Right*: ordering potential $(k_B T/2)c_{cc}(r)$ extracted from the experimental data [4]

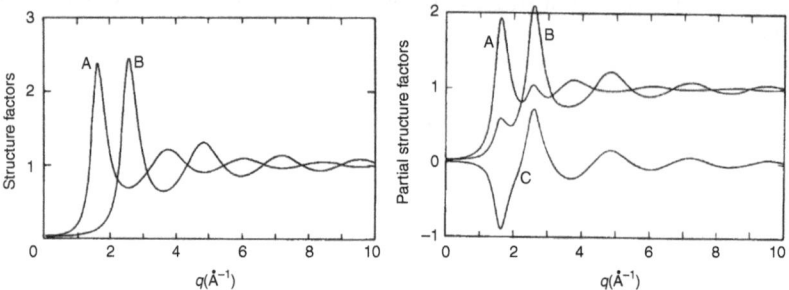

Fig. 3.8 *Left*: A $S_{cc}(q)$ and B $S_{\rho\rho}(q)$ calculated in MSA from (3.76) [4]. *Right*: A $S_{LiLi}(q)$, B $S_{LiPb}(q)$, C $S_{PbPb}(q)$ from the same calculation

$\bar{b} = c_A b_A + c_B b_B = 0$. In this case one can obtain the function $c_{cc}(q)$ from the measured $S_{cc}(q)$. via the relation (3.74). Its Fourier transform is just $c_{cc}(r)$, which is—if one believes in the RPA/MSA relation (3.72)—proportional to the ordering potential $\phi_{ord}(r)$ for r larger than the hard-core radius d.

As can be seen from Fig. 3.7, the part for $r > 3$ Å is temperature independent and can be fitted by an *attracting* ordering potential of the form (3.75) with $\lambda = 1.1$ Å^{-1}. In the paper of Copestake et al. [4] an Ansatz was now made for the interatomic potentials just as in the case of the metal-halogen mixtures (3.56)

$$\phi_{ij}(r) = \phi_{ij}^{(0)}(r) + Q_i Q_j e^2 \frac{e^{-\lambda r}}{r} \qquad i = A, B \qquad (3.76)$$

The idea is that Pb is much more electronegative than Li and therefore acquires a negative charge $Q_B e$. This leads to a strong attractive ordering potential of the form (3.75). If the global charge neutrality condition $\bar{Q} = c_{Li} Q_{Li} + c_{Pb} Q_{Pb} = 0$ is imposed the "regular solution conditions" (3.69) and (3.70) hold.

For this potential a MSA calculation with $d = 3$ Å has been performed [4], which enabled to calculate $S_{cc}(q)$, $S_{\rho\rho}(q)$ and from that the three partial structure factors (Fig. 3.8).

In this way one was able to gain a lot of insight into the structural properties of this compound-forming alloy system.

3.12.2 Critical Scattering in Mixtures with Demixing Transition

We consider again a regular solution with a free enthalpy of mixing of the form

$$\frac{1}{N}\Delta G = c(1-c)\omega + k_B T \left[c_A \ln c_A + c_B \ln c_B\right] \tag{3.77}$$

which corresponds to

$$\frac{1}{S_{cc}(0)}\frac{1}{Nk_BT}\frac{\partial^2 \Delta G}{\partial c^2} = \frac{1}{k_BT}g_{cc} = \frac{1}{c_A c_B} - \frac{2\omega}{k_BT} \tag{3.78}$$

We now make the RPA Ansatz

$$\frac{1}{S_{cc}(q)} = \frac{1}{c_A c_B} - \frac{2\phi_{\text{ord}}(q)}{k_BT} \tag{3.79}$$

We expect that the ordering potential—just as all other direct correlation functions—decay on an atomic scale, so we make the same Ansatz for $\phi_{\text{ord}}(r)$ as in the Li$_4$Pb case

$$\phi_{\text{ord}}(r) \propto \frac{1}{r}e^{-\lambda r} \qquad \Leftrightarrow \qquad \phi_{\text{ord}}(q) = \frac{\phi_{\text{ord}}(0)}{1+\frac{q^2}{\lambda^2}} = \frac{\omega}{1+\frac{q^2}{\lambda^2}}, \tag{3.80}$$

where λ is supposed to be of the order of an interatomic distance. For small q we can expand $\phi_{\text{ord}}(q)$

$$\phi_{\text{ord}}(q) \overset{q\to 0}{=} \omega\left(1 - \frac{q^2}{\lambda^2}\right) \tag{3.81}$$

We are now interested in the properties of the structure factor $S_{cc}(q)$ for small q near the critical point. Inserting (3.81) into (3.79) and take $c_A = c_B = \frac{1}{2}$ we obtain

$$S_{cc}(q) = \frac{1}{4 - \frac{2\omega}{k_BT} + \frac{2\omega}{k_BT}q^2/\lambda^2} = \frac{k_BT}{2\omega}\frac{\lambda^2}{\xi^{-2} + q^2} \tag{3.82}$$

which corresponds to

$$h_{cc}(r) \propto \frac{1}{r}e^{-r/\xi} \tag{3.83}$$

Fig. 3.9 Flory–Huggins
model of a polymer as a
random walk on a lattice

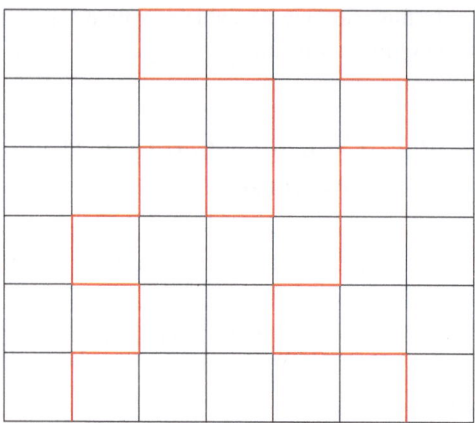

ξ is the correlation length which has the temperature dependence

$$\xi(T) = \frac{\lambda^{-1}}{\sqrt{\frac{2k_BT}{\omega} - 1}} = \frac{\lambda^{-1}}{\sqrt{\frac{2k_BT}{\omega}(T - T_c)}} \tag{3.84}$$

In contrast to the microscopic quantity λ^{-1} the correlation length ξ becomes macroscopically large near the critical point. In the regime $T > T_c$ one can observe strong *small-angle scattering* as a precursor of the demixing transition.

3.13   Solutions of Polymers

We now consider N_p polymer molecules consisting of N segments immersed in a solvent consisting of N_s molecules. The total number of molecules[1] is $N_m = N_s + N_p$. Following Flory [5] and Huggins [8] we imagine that we arrange the monomers and the solvent molecules on a simple cubic lattice (volume V, lattice constant a), (Fig. 3.9) which provides $N_V = V/a^3 = N N_p + N_s$ sites. With respect to the concentrations $c_p = c = N_p/N_m$, $c_s = 1 - c = N_s/N_m$ the Entropy of mixing takes the form

$$\Delta S(c) = -N_m k_B [c \ln c + (1 - c) \ln(1 - c)]. \tag{3.85}$$

Instead of working with the polymer concentration we want to work with the *volume fraction* ϕ occupied by the polymer. This number is given by

---

[1]This number was called N up to now. As in the polymer literature N is reserved to the number of segments, we re-define $N \to N_m$.

$$\phi = N\frac{N_p}{N_V} = Nc_p\frac{N_m}{N_v} .$$ (3.86)

Consequently we have for c and $1 - c$

$$c = \frac{\phi}{N}\frac{N_V}{N_m} \qquad 1 - c = (1 - \phi)\frac{N_V}{N_m}$$ (3.87)

So that we obtain

$$\Delta S_c = -N_V k_B \left[\frac{\phi}{N} \ln \frac{\phi N_V}{N N_m} + (1 - \phi) \ln(1 - \phi)\frac{N_V}{N_m} \right]$$

$$= -N_V k_B \left[\frac{\phi}{N} \ln \phi + (1 - \phi) \ln(1 - \phi) + \frac{\phi}{N} \ln \frac{N_V}{N N_m} + (1 - \phi) \ln \frac{N_V}{N_m} \right] .$$ (3.88)

With respect to the variable ϕ we obtain the following entropy of mixing per lattice site

$$\Delta S = \frac{1}{N_V} [\Delta S_c(\phi) - \phi \Delta S_c(\phi = 1) - (1 - \phi)\Delta S_c(\phi = 0)]$$

$$= -k_B \left[\frac{\phi}{N} \ln \phi + (1 - \phi) \ln(1 - \phi) \right] .$$ (3.89)

Now we add an enthalpy term of the same form as that in the theory of regular solution, except that we now work with the variable ϕ:

$$\Delta H = \phi(1 - \phi)\epsilon = \phi(1 - \phi)\chi k_B T$$ (3.90)

Here ϵ is the energy difference

$$\epsilon = z \left[\epsilon_{ps} - \frac{1}{2}(\epsilon_{pp} + \epsilon_{ss}) \right]$$ (3.91)

between $p-s$ nearest-neighbors and the averaged $p-p$ and $s-s$ nearest-neighbors. z is the lattice coordination number (i.e., $z = 6$). So we obtain the *Flory–Huggins expression* for the free enthalpy per site and $k_B T$

$$\Delta g = \frac{1}{k_B T}\Delta G = \phi(1 - \phi)\chi + \frac{\phi}{N} \ln \phi + (1 - \phi) \ln(1 - \phi)$$ (3.92)

Of course this model embodies a very crude approximation to the reality, as the different possible polymer chain conformations are completely neglected. However, it has turned out that despite the model is more than 50 years old it still serves as

Fig. 3.10 Phase diagram for
the Flory–Huggins model.
Straight lines are the
spinodals, Eq. (3.93), the
dotted curve is the
coexistence curve, Eq. (3.99),
for $N \to \infty$

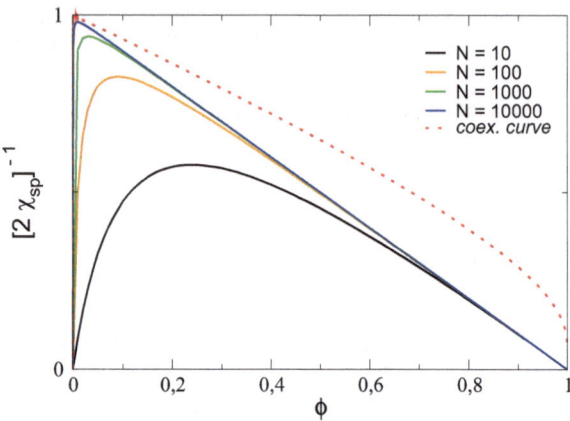

a starting point for discussing the thermodynamics of polymer solutions. For the
spinodal value of χ we obtain (Fig. 3.10)

$$2\chi_{sp} = \frac{1}{N\phi} + \frac{1}{1-\phi} \qquad (3.93)$$

If we seek the minimum of this curve we obtain for the critical concentration

$$\phi_c = \frac{\sqrt{N}-1}{N-1} \approx \frac{1}{\sqrt{N}} \qquad (3.94)$$

We see that for large N the critical point becomes situated at very dilute volume
fractions, and the critical temperature approaches the so-called Θ limit:

$$\lim_{N\to\infty} \chi_c = \lim_{N\to\infty} \frac{\epsilon}{k_B T_c} = \frac{\epsilon}{k_B \Theta} = 2 \qquad (3.95)$$

From the equality of the chemical potentials of the concentrated and the
dilute phase we have for the equilibrium volume fractions $\phi_{1,2}$: (double-tangent
construction)

$$\frac{\Delta g_1 - \Delta g_2}{\phi_1 - \phi_2} = \left.\frac{\partial \Delta g}{\partial \phi}\right|_{\phi_1} = \left.\frac{\partial \Delta g}{\partial \phi}\right|_{\phi_2} \qquad (3.96)$$

We know that the equilibrium volume fractions for the concentrated and the
dilute phase must be situated outside the spinodal curve. As for large N there is
not much of an interval left for the concentrated volume fraction, we conclude that

$$\lim_{N\to\infty} \phi_1 = 0 , \qquad (3.97)$$

from which follows in the limit $N \to \infty$:

$$0 = \phi_2 \left.\frac{\partial \Delta g}{\partial \phi}\right|_{\phi_2} - \Delta g_2 = -\phi_2^2 \chi + \phi_2 \left(\frac{1}{N} - 1\right) + \ln \frac{1}{1 - \phi_2}$$

$$\approx -\phi_2^2 \chi + \ln \frac{1}{1 - \phi_2} - \phi_2 . \tag{3.98}$$

This leads to a coexistence curve of the form

$$\frac{1}{2\chi} = \frac{\phi_2^2}{2} \frac{1}{\ln \frac{1}{1-\phi_2} - \phi_2} = \frac{\phi_2^2}{2} \frac{1}{\frac{1}{2}\phi_2^2 + \frac{1}{3}\phi_2^3 + \frac{1}{4}\phi_2^4 + \cdots} \tag{3.99}$$

We see that—according to the Flory–Huggins RPA theory—in the phase-separated state the concentrated phase contains still a lot of solvent molecules, whereas the dilute phase is entirely made up of polymer solute molecules.

References

1. Bhatia, A.B.: In: Evans, R., Greenwood, D.A. (eds.) Liquid Metals 1976, p. 21. The Institute of Physics, London (1976)
2. Bhatia, A.B., Thornton, D.E.: Phys. Rev. B **2**(8), 3004 (1970)
3. Bredig, M.A.: Wiley Interscience, New York (1964)
4. Copestake, A.P., Evans, R., Ruppersberg, H., Schirmacher, W.: J. Phys. F Met. Phys. **13**, 1993 (1983)
5. Flory, P.J.: Chem. Phys. **9**, 660 (1941)
6. Holzhey, C., Schirmacher, W.: J. Phys. **46**, C8 (1985)
7. Holzhey, C., Schirmacher, W.: Z. Phys. Chem. (Neue Folge) **156**, 163 (1988)
8. Huggins, H.L.: Chem. Phys. **9**, 440 (1941)
9. Lebowitz, J. L., Phys. Rev. A **133**, 89(5) (1964)
10. Ruppersberg, H., Reiter, H.: J. Phys. F Met. Phys. **12**, 1311 (1982)

Introduce following designation $X = \lambda t$

$$\psi(X) = \frac{1}{\sqrt{\pi}} \int_0^X e^{-z^2}\,dz$$

for such a hyperbolic case the form

$$\frac{1}{\sqrt{\pi}} \int \ldots$$

Chapter 4
Random Walk and Diffusion

4.1 Einstein's Theory of Brownian Motion

In his famous article (one of four in 1905) *"Über die von der molekularkineti-schen Theorie der Wärme geforderte Bewegung von in ruhenden Flüssigkeiten sus-pendierten Teilchen"*,[1] Ann. Phys. (Leipzig) **17**, 549, Einstein considers small solid particles suspended in a solvent. He argues that from the standpoint of the kinetic theory of heat such particles should behave just in the same way as a solute and give—in particular—rise to an osmotic pressure felt by a membrane which is permeable for the solvent and non-permeable for the particle. This pressure p^* stems from the collisions with the membrane, just as the gas pressure stems from the collision of the gas molecules with the wall of the gas container (Fig. 4.1).

He then considers the influence of virtual changes in the coordinates δx of the suspended particles, the density of which is $\rho(x) = N/V$ inside of a box of length ℓ and cross-section 1. In thermodynamic equilibrium the free energy $F = E - TS$ should not be affected by δx as the change in entropy ΔS due to Δx should just counterbalance that in the internal energy ΔE:

$$\Delta E - T\Delta S = 0. \tag{4.1}$$

In terms of a force F on a particle which suffers a virtual coordinate displacement δx ΔE is given by

$$\Delta E = -\int_0^\ell dx \rho(x) F \delta x \tag{4.2}$$

---

[1] On the motion of particles suspended in liquids at rest, which has been postulate by the kinetic theory of heat.

W. Schirmacher, *Theory of Liquids and Other Disordered Media*, Lecture Notes in Physics 887, DOI 10.1007/978-3-319-06950-0_4,
© Springer International Publishing Switzerland 2015

Fig. 4.1 Einstein's box of
length ℓ and of
cross-section 1

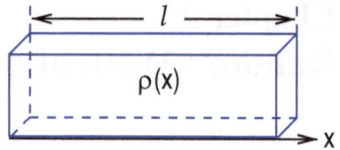

The entropy S due to the distribution density $\rho(x)$ was calculated by him
assuming—as done in the theory of dilute solutions—that a *partial* gas equation
holds for the ensemble of the particles:

$$p^* = \frac{R}{\mathcal{N}_{\text{Avo}}}\rho T \tag{4.3}$$

so that

$$S = \int_0^\ell dx \frac{p^*}{T} = \frac{R}{\mathcal{N}_{\text{Avo}}} \int_0^\ell dx \rho(x) \tag{4.4}$$

If there is now a coordinate displacement δx we have from a Taylor expansion of
$\rho(x)$ into the integrand and performs an integration by part to obtain

$$\Delta S = \frac{R}{\mathcal{N}_{\text{Avo}}} \int_0^\ell dx \rho(x) \frac{\partial}{\partial x}\delta x , = -\frac{R}{\mathcal{N}_{\text{Avo}}} \int_0^\ell dx \frac{\partial \rho(x)}{\partial x}\delta x , \tag{4.5}$$

where the second equality follows from an integration by part. As the variations δx
are supposed to be arbitrary we can state

$$F\rho - \frac{RT}{\mathcal{N}_{\text{Avo}}}\frac{\partial \rho}{\partial x} = F\rho - \frac{\partial p^*}{\partial x} = 0 \tag{4.6}$$

Einstein now turns his attention from the thermodynamic equilibrium to the *dynamic*
equilibrium of microscopic currents. Due to the force F the particle (assumed to be
spherical with radius r) acquires a velocity due to Stokes friction

$$v = \frac{F}{6\pi \eta r} \tag{4.7}$$

where η is the viscosity. Inserting Fick's law for the diffusion current

$$j = \rho v = -D\frac{\partial \rho}{\partial x} \tag{4.8}$$

we obtain for the dynamic equilibrium of currents

$$\rho \frac{F}{6\pi \eta r} - D\frac{\partial \rho}{\partial x} = 0 \tag{4.9}$$

Combining (4.6) and (4.9) we obtain the famous Einstein relation between the diffusion coefficient and the viscosity

$$D = \frac{1}{\mathcal{N}_{\text{Avo}}} \frac{RT}{6\pi \eta r} \tag{4.10}$$

In this paragraph we did not use the Boltzmann constant $k_B = R/\mathcal{N}_{\text{Avo}}$ for the following reason: In 1905 neither Avogadro's number was known nor was the atomic hypothesis acknowledged by all of the physical community. Einstein suggested to perform measurements of the diffusion coefficient of Brownian particles to estimate $\mathcal{N}_{\text{Avo}}$.

4.2 Diffusion Equation and Mean-Square Displacement

We now deviate from Einstein's original manuscript and consider the diffusion process of Brownian particles as it is presented nowadays in textbooks.

We first consider—as above—only motions in the x-direction. We start with the continuity equation for the time dependence of the particle density $\rho(x,t)$ and the current density $j(x,t) = \rho v$, which holds due to the conservation of the total number of Brownian particles:

$$\frac{\partial \rho}{\partial t} + \frac{\partial j}{\partial x} = 0, \tag{4.11}$$

if we then insert Fick's law (4.8) we obtain the diffusion equation

$$\frac{\partial \rho}{\partial t} - D \frac{\partial^2 \rho}{\partial x^2} = 0. \tag{4.12}$$

Going over to a spatial Fourier transform

$$\rho(k,t) = \int_{-\infty}^{\infty} dx \, e^{ikx} \rho(x,t) \tag{4.13}$$

and a temporal Laplace transform

$$\rho(k,s) = \int_{0}^{\infty} dt \, e^{-st} \rho(k,t) \tag{4.14}$$

we obtain from (4.12)

$$s\rho(k,s) - \rho(k,t=0) + Dk^2\rho(k,s) = 0 \tag{4.15}$$

with the solution

$$p(k,s) = \frac{p(k,t=0)}{s+Dk^2} = p(k,t=0)G(k,s) \tag{4.16}$$

where we have introduced the special solution $G(k,t)$ (Green's function), which has the initial condition

$$G(k,t=0) = 1 \tag{4.17}$$

corresponding to

$$G(x,x_0t=0) = \delta(x-x_0) \tag{4.18}$$

where x_0 is the initial position of the random walker. So this function obeys the differential equation of motion[2]

$$sG(x,x_0,s) - D\frac{\partial^2}{\partial x^2}G(x,x_0,s) = \delta(x-x_0) \tag{4.19}$$

The Green's function $G(x,t)$ can be interpreted as the probability density of a Brownian particle which started his journey at $t=0$ at the origin $x=0$ (see next section). The back transforms of $G(k,p)$ are

$$G(k,t) = e^{-Dk^2t} \tag{4.20}$$

and

$$G(x,t) = \frac{1}{\sqrt{4\pi Dt}}e^{-x^2/4Dt} \tag{4.21}$$

An important quantity is the *mean square distance* walked by the Brownian particle at a certain time t. It can be calculated from the function $G(x,t)$ as

$$\langle x^2(t)\rangle = \int_{-\infty}^{\infty} dx\, x^2\, G(x,t) \tag{4.22a}$$

$$= -\frac{\partial^2}{\partial k^2}\int_{-\infty}^{\infty} dx\, e^{ikx}\, G(x,t)\Big|_{k=0} \tag{4.22b}$$

---

[2]A function obeying a differential equation with a delta-function inhomogeneity like Eq. (4.19) is called a *Green's function*, because it is helpful to solve an inhomogeneous differential equation with an arbitrary inhomogeneity $j(x)$. A theorem then states that a special solution of the inhomogeneous equation is given by the convolution of the inhomogeneity and the Green's function.

$$= -\frac{\partial^2}{\partial k^2} e^{-Dk^2 t}\Big|_{k=0} \tag{4.22c}$$

$$= 2Dt \tag{4.22d}$$

This is the second formula in the theory of Brownian motion which carries Einstein's name. It states that the distance a Brownian particle moves away on the average from its starting point grows with the square-root of time.

The diffusion equation and its solution are easily generalized to the three-dimensional case. The diffusion equation reads

$$\frac{\partial \rho(\mathbf{r}, t)}{\partial t} - D\nabla^2 \rho(\mathbf{r}, t) = 0, \tag{4.23}$$

with solution

$$G(\mathbf{k}, t) = \int_{-\infty}^{\infty} d^3 r \, e^{i\mathbf{k}\cdot\mathbf{r}} G(\mathbf{r}, t) = e^{-Dk^2 t}, \tag{4.24}$$

subject to the initial condition $G(\mathbf{k}, t = 0) = 1 \Leftrightarrow G(\mathbf{k}, t = 0) = \delta(\mathbf{r})$. We now have $k^2 = |\mathbf{k}|^2 = k_x^2 + k_y^2 + k_z^2$. The solution in $\mathbf{r}$ space has the form

$$G(\mathbf{r}, t) = \left[\frac{1}{\sqrt{4\pi Dt}}\right]^3 e^{-r^2/4Dt} \tag{4.25}$$

with $r^2 = |\mathbf{r}|^2 = x^2 + y^2 + z^2$. For the three-dimensional mean-square distance we obtain the Einstein relation

$$\langle r^2(t) \rangle = \int_{-\infty}^{\infty} d^3 r \, \left(x^2 + y^2 + z^2\right) G(\mathbf{r}, t) \tag{4.26a}$$

$$= -\left(\frac{\partial^2}{\partial k_x^2} + \frac{\partial^2}{\partial k_y^2} + \frac{\partial^2}{\partial k_z^2}\right) \int_{-\infty}^{\infty} d^3 r \, e^{i\mathbf{k}\cdot\mathbf{r}} G(\mathbf{r}, t)\Big|_{k_x=k_y=k_z=0} \tag{4.26b}$$

$$= -\left(\frac{\partial^2}{\partial k_x^2} + \frac{\partial^2}{\partial k_y^2} + \frac{\partial^2}{\partial k_z^2}\right) e^{-Dk^2 t}\Big|_{k_x=k_y=k_z=0} \tag{4.26c}$$

$$= 6Dt \tag{4.26d}$$

4.3   Random Walk on a Lattice

The motion of a Brownian particle can be visualized by that of a "random walker", e.g., by a drunken person, who changes its direction at random after every step (Fig. 4.2). The statistics of such a motion can be easily worked out on a lattice. We

Fig. 4.2 A random walker on
a two-dimensional lattice

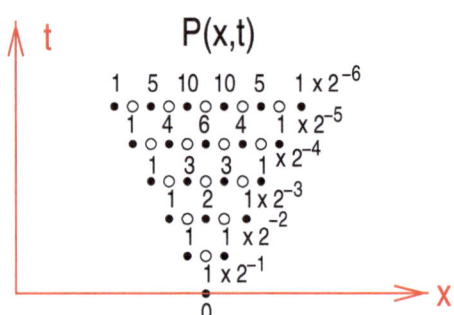

Fig. 4.3 Pascal's triangle for
a 1d random walk. The
number of ways a time-space
point can be reached from the
bottom is given by Pascal's
algorithm, i.e., each number
is the sum of the two numbers
underneath

start with this on a one-dimensional array of points with lattice constant a, the time
steps are called τ. The probabilities $P(x = a, t = \tau)$ and $P(x = -a, t = \tau)$ are
$1/2$, those for one time step for $|x| > a$ are zero. The non-zero probabilities for the
vth time step are $(1/2)^v$ times the number of ways one can reach the site $x_n = na$
on the triangle (Fig. 4.3).

This number increases from 1 at the maximum distance $|x|_{max} = va$ with $k =$
$(|x|_{max} - |x|)/a$ as $\binom{n}{k}$, which can be represented as

$$P(x_n, v\tau) = \left(\frac{1}{2}\right)^v \binom{v}{k} \qquad k = \left[\frac{1}{2}(n + v)\right], \tag{4.27}$$

where $[v]$ is the smallest integer κ with $\kappa \geq v$. It is worth wile to note that at an
even/odd time step v only even/odd random walk sites x_n can be reached (Fig. 4.3).

We consider now the recursion formula for the binomial coefficients

$$\binom{n+1}{k} = \binom{n}{k} + \binom{n}{k-1} \tag{4.28}$$

We re-write this recursion formula with the help of the probabilities $P(nx, v\tau)$:

$$P(x_n, t + \tau) = \left(\frac{1}{2}\right)^{v+1} \binom{v+1}{k_{n,v+1}} = \frac{1}{2}P(x_{n+1}, t) + \frac{1}{2}P(x_{n-1}, t) \tag{4.29}$$

This can be re-written as

$$\frac{P(x_n, t + \tau) - P(x, t)}{\tau} = D\left[\frac{P(x_n + a) + P(x_n - a, t) - 2P(x, t)}{a^2}\right] \quad (4.30)$$

with

$$D = \frac{a^2}{2\tau} \quad (4.31)$$

We take now the double limit $\tau \to 0$ and $a \to 0$ keeping the ratio $D = a^2/2\tau$ fixed. This leads to

$$\frac{\partial P(x, t)}{\partial t} = D\frac{\partial^2}{\partial x^2}P(x, t) \quad (4.32)$$

This is just the diffusion equation for the Brownian motion encountered in the previous chapter. So the Function $G(x, t)$ is just the continuum limit of the random walk probability.

4.3.1 Master Equation

We can establish another useful limit of the difference equation (4.29). We write again

$$\frac{P(x_n, t + \tau) - P(x_n, t)}{\tau} = \frac{1}{2\tau}[P(x_{n+1}, t) + P(x_{n-1}, t) - 2P(x_n, t)]$$

$$= \sum_{\ell=\pm 1} W_{n\ell}[P(x_\ell, t) - P(x_n, t)] \quad (4.33)$$

with $W_{n\ell} = 1/2\tau \equiv W$. Here we have introduced the transition rate per unit time $W_{n\ell} = W_{\ell n}$ for the walker to make a transition from n to ℓ. For times $t \gg \tau$ this can be written as

$$\frac{d}{dt}P(x_n, t) = W\sum_{\ell=\pm 1}[P(x_\ell, t) - P(x_n, t)] \quad (4.34)$$

This equation is called the *master equation* for the random walk probability $P(x_n, t)$. As initial condition we have

$$P(x_n, t = 0) = \delta_{n,0} \quad (4.35)$$

(4.34) can readily be solved by taking the time-Fourier transform of $P(x_n, t)$, $P(x_n, \omega)$ and applying Bloch's theorem

$$P(x, \omega) = e^{ikx} u_k(x, \omega) \tag{4.36}$$

where $u_k(x, \omega)$ has the periodicity of the lattice. So we have for $u_k(\omega) = u_k(0, \omega)$

$$i\omega u_k(\omega) = 2W[\cos kx - 1]u_k(\omega) \tag{4.37}$$

For the Laplace transform $u_k(s) = \int_0^\infty dt\, u_k(t)e^{-st}$ we obtain

$$su_k(s) - 1 = 2W[\cos kx - 1]u_k(s) \tag{4.38}$$

with the solution

$$u_k(s) = \frac{1}{s + 2W(1 - \cos ka)} \tag{4.39}$$

We now recognize that $u_k(s)$ has the property

$$\lim_{k \to 0} u_k(s) = \frac{1}{s + Wa^2k^2} \tag{4.40}$$

We see that we can identify $u_k(s)$ in this limit with the solution of the diffusion equation $G(k, s)$ with diffusion coefficient $D = a^2W = a^2/2\tau$ in agreement with the result (4.32).

The three-dimensional generalization of (4.39) is

$$u_{\mathbf{k}}(s) = \frac{1}{s + Wf(k)} = G(k, s) \tag{4.41}$$

with the dimensionless "band structure" of the simple-cubic lattice

$$f(k) = 6 - 2\cos k_x a + 2\cos k_y a + 2\cos k_z a \tag{4.42}$$

In the $k \to 0$ limit one obtains (4.16) for $G(k, s)$ with $D = a^2W = a^2/2\tau$.

4.4   Disordered Lattice and the Coherent-Potential Approximation (CPA)

An interesting situation arises, (which is quite relevant for diffusion in glasses or other soft matter) when the random walk is affected by the presence of disorder, which can be either that some bonds with concentration $1 - p$ are taken out of the lattice or one has a statistical distribution of transition rates $W_{n\ell}$. The equation of motion of this system is (4.34) but now in three dimensions (simple-cubic lattice) and with different $W_{n\ell}$'s:

Fig. 4.4 Effective-Medium procedure for the two-site coherent-potential approximation (CPA)

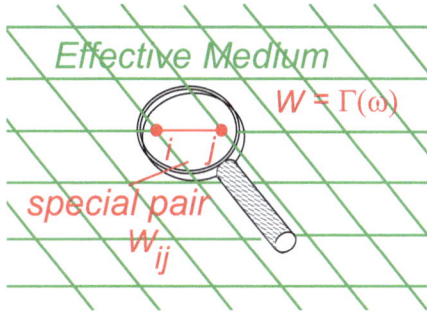

$$\frac{d}{dt} P(\mathbf{r}_n, t) = \sum_{\substack{\ell \\ \text{n.N.}}} W_{n\ell} \left[P(\mathbf{r}_\ell, t) - P(\mathbf{r}_n, t) \right] \tag{4.43}$$

If we set the left side of Eq. (4.43) equal to zero, we obtain the steady-state condition

$$\sum_{\substack{\ell \\ \text{n.N.}}} W_{n\ell} \left[P(x_\ell) - P(x_n) \right] = 0 , \tag{4.44}$$

which can be interpreted as Kirchhoff's equations for a network with nodes at the sites $\mathbf{r}_n$ with ingoing local currents $P(\mathbf{r}_\ell)$ and outgoing currents $P(\mathbf{r}_n)$. This means that we can map a random-walk problem to a network problem, which is well studied in electrical engineering [7].

Returning to the discussion of Eq. (4.43), we define now a lattice Green's function $\mathcal{G}_{ij}(t, t')$ which obeys the equation

$$\frac{d}{dt} \mathcal{G}_{ij}(t, t') + \sum_\ell W_{i\ell}(\mathcal{G}_{ij} - \mathcal{G}_{\ell j}) = \delta_{ij} \delta(t - t') \tag{4.45}$$

Defining a "Hamiltonian Matrix"

$$\mathcal{H}_{ij} = \begin{cases} -\sum_{\ell \neq 1} W_{i\ell} & i = j \\[2mm] W_{ij} & i \neq j \end{cases} \tag{4.46}$$

and going into frequency space with a complex frequency variable $s = -i\omega + \epsilon$ (where the very small number ϵ has to be included for mathematical reasons), we obtain the following matrix equation for the Green matrix $< i|\mathcal{G}(\omega)|j> = \mathcal{G}_{ij}(\omega)$

$$(s - \mathcal{H})\mathcal{G} = 1 \tag{4.47}$$

One of the most powerful mean-field theories of disorder is the *coherent-potential approximation, CPA*. The CPA is derived as follows:

We invent an *effective medium*,[3] which is not disordered (i.e., it has the cubic symmetry), but the force constants are frequency dependent:

$$W_{ij}^{\text{eff}}(s) = \Gamma(s).\tag{4.48}$$

The effective-medium Hamiltonian is

$$H_{ij}^{\text{eff}} = \begin{cases} -\sum_j W_{ij}^{\text{eff}}(s) = -Z\Gamma(s) & i = j \\[2mm] W_{ij}^{\text{eff}}(s) = \Gamma(s) & i \neq j \end{cases}\tag{4.49}$$

where $Z(=6)$ is the number of nearest neighbors (coordination number). The Green's function of the effective medium obeys the equation of motion

$$sG_{ij} - \delta_{ij} = Z\Gamma(s)(G_{\ell j} - G_{ij}) \qquad \ell \text{ arbitrary neighboring site}\tag{4.50}$$

As in the Lorentz theory of dielectric polarizability [2] we now consider a particular region of the effective medium with the reading glass (Fig. 4.4), which just contains a pair of sites (i_0, j_0). Inside this region we replace the effective-medium force constant $\Gamma(s)$ by the actual one $W_{i_0 j_0}$ so that we obtain a "perturbation" $v_{i_0 j_0}(s) = W_{i_0 j_0} - \Gamma(s)$. The corresponding perturbing Hamiltonian matrix V has four non-zero entries, namely $V_{i_0 i_0}$, $V_{j_0 j_0}$, $V_{i_0 j_0}$, and $V_{j_0 i_0}$. In the $i_0 j_0$ subspace we have

$$V = \begin{pmatrix} -v_{i_0 j_0}(s) & v_{i_0 j_0}(s) \\[2mm] v_{i_0 j_0}(s) & -v_{i_0 j_0}(\tilde{s}) \end{pmatrix}\tag{4.51}$$

We now demand that introducing this perturbation should have *on the average* no influence on the effective medium which is equivalent to demanding that the Green's function of the effective medium should be equal to the configurationally averaged Green's function of the disordered system. One can again reformulate this postulate in demanding that the averaged *T-matrix*

$$\langle T \rangle = \left\langle \frac{V}{1 - VG} \right\rangle\tag{4.52}$$

should vanish. Working out the 2×2 inverse and using (4.50) with $i = j$ we obtain the following CPA self-consistent equation for $\Gamma(s)$

---

[3]The first version of the CPA effective-medium approach has been devised by Bruggeman [3] for calculating the dielectric constant of composite dielectrica.

$$\left\langle \frac{W - \Gamma(s)}{1 + (W - \Gamma(s))\frac{2}{Z} \underbrace{\frac{1}{\Gamma(s)}(1 - sG_{ii}(s))}_{\Lambda(s)}} \right\rangle = 0 \tag{4.53}$$

with the local Green's function

$$G_{ii}(s) = \sum_{\mathbf{k} \in BZ} \frac{1}{s + \Gamma(s)f(\mathbf{k})} \tag{4.54}$$

Equation (4.53) can be reformulated as

$$\Gamma(s) = \left\langle \frac{W}{1 + (W - \Gamma(s))\frac{2}{Z}\Lambda(s)} \right\rangle \tag{4.55}$$

and also the function $\Lambda(s)$ can be simplified:

$$\Lambda(s) = \frac{1}{\Gamma(s)}(1 - sG_{ii}(s)) = \sum_{\mathbf{k} \in BZ} \frac{f(\mathbf{k})}{s + \Gamma(s)f(\mathbf{k})} \tag{4.56}$$

Because $\sum_{\mathbf{k} \in BZ} 1 = 1$ one obtains in the $s \to 0$ limit

$$\Lambda(0) = \frac{1}{\Gamma(0)}, \tag{4.57}$$

so that Eq. (4.55) in the dc limit reads

$$\Gamma(0) = \left\langle \frac{W}{1 + \frac{2}{Z}\left(\frac{W}{\Gamma(0)} - 1\right)} \right\rangle$$

$$= \frac{Z}{2}\Gamma(0)\left\langle \frac{1}{\left(\frac{Z}{2} - 1\right)\frac{\Gamma(0)}{W} + 1} \right\rangle. \tag{4.58}$$

4.4.1 Percolating Lattice

We assume now a distribution of local diffusivities of the form

$$P(D_i) = p\delta(W - W_0) + (1 - p)\delta(W) \tag{4.59}$$

where p is the volume fraction in which the transition rate $W_{i_0 j_0}$ is non-zero. If we just consider the statistics of the bonds, which are taken out of the lattice with

concentration $1 - p$, this constitutes the *percolation model*, which we are going
to discuss shortly in the chapter on fractals. Inserting the distribution given by
Eq. (4.59) into Eq. (4.58) one obtains

$$\Gamma(0)\frac{2}{Z} = \Gamma(0)\frac{1}{\left(\frac{Z}{2} - 1\right)\frac{\Gamma(0)}{W_0} + 1} \tag{4.60}$$

which has, like Eq. (4.58) always the trivial solution $\Gamma(0) = 0$. For the case $\Gamma(0) \neq 0$
we obtain from Eq. (4.60)

$$\Gamma(0) = \frac{2}{3}W_0(p - p_c)/(1 - p_c) \tag{4.61}$$

At the critical value $p_c = \frac{2}{Z} \overset{Z=6}{=} \frac{1}{3}$ the transition rate $\Gamma(0)$ of the effective medium,
and hence the diffusivity $D = W_0 a^2$ vanishes. Beyond this value the trivial solution
$W_0 = D = 0$ takes over. One easily verifies that right at $p = p_c$ the frequency-
dependent diffusivity (see Appendix C) varies with frequency as $D(s) = a^2\Gamma(s) \propto$
$s^{1/2}$, which implies that the mean square displacement grows with time as $t^{1/2}$ at
criticality.

4.4.2   Continuum Limit: Activated Diffusion with Disorder

Now we can again take the continuum limit in order to obtain a continuum version
of the CPA [8]. We introduce local, spatially fluctuating diffusivities $D(\mathbf{r}_{i_0 j_0}) =
a^2 W r_{i_0 j_0}$, where $\mathbf{r}_{i_0 j_0}$ denotes the center of gravity of the bond i_0, j_0, an effective-
medium diffusivity $D(s)$ and a susceptibility function $\chi(s) = \Lambda(s)/a^2$ and let $a \to$
0. We further replace the BZ summation by the summation over $\mathbf{k}$ space up to the
Debye cutoff [1] $k_F = \sqrt[3]{6\pi^2 N/V j}$, i.e

$$\sum_{\mathbf{k}\in BZ} \to \frac{1}{N}\sum_{|\mathbf{k}|<k_F} = \frac{3}{k_F^3}\int_0^{k_F} k^2 dk$$

In the continuum limit the CPA now takes the form

$$D(s) = \left\langle\frac{D(\mathbf{r})}{1 + (D(\mathbf{r}) - D(s))\frac{1}{3}\chi(s)}\right\rangle = 0 \tag{4.62}$$

with

$$\chi(z) = \frac{1}{N}\sum_{|\mathbf{k}|<k_F}\frac{k^2}{-z^2 + D(s)k^2} \tag{4.63}$$

We now consider a diffusing particle, which performs activated jumps over barriers with a given distribution of activation energies $P(E(\mathbf{r}))$. ("random barrier model" [4]). So we write

$$D(\mathbf{r}) = D_0 e^{E(\mathbf{r})/k_B T} \tag{4.64}$$

and parametrize the *dc* effective-medium diffusivity as $(\frac{Z}{2} - 1)D(0) = D_0 e^{E_a/k_B T}$. Then Eq. (4.58) takes the form

$$\frac{2}{Z} = \int dE P(E) \frac{1}{e^{(E-E_a)/k_B T} + 1} \tag{4.65}$$

The function to be integrated over together with the distribution function is the Fermi function, which strongly suppresses the regime $E > E_a$. In the low-temperature limit $T \to 0$ the Fermi function becomes a step function and we obtain

$$\frac{2}{Z} = p_c = \frac{1}{3} = \int_0^{E_a} dE P(E) \tag{4.66}$$

which means that the parameter E_a becomes temperature independent. We can understand this result with the *percolation construction* [5] for the random-barrier problem. As we noted already, the problem of finding the *dc* conductivity corresponding to the hopping process described by the equation of motion (4.43) can be mapped to the problem of finding the global conductance of a conductance network [7]. We now sort the conductances $g_i \propto D(\mathbf{r}_i)$ by their magnitude, which is determined by the activation energy $E(\mathbf{r}_i)$ and remove first all conductances from the network. We then start soldering in the largest conductances with an activation energy smaller than $E_{\text{threshold}}$. If $E_{\text{threshold}}$ is small enough this will result in a set of isolated clusters of connected nodes. If we now increase $E_{\text{threshold}}$ the clusters of connected nodes will grow and constitute a percolation scenario. At a certain value E_{critical} percolation happens, and the network will essentially have the conductance $g \propto e^{-E_{\text{critical}}/k_B T}$, because all other (much larger) conductances are essentially shortcuts. We can visualize this by saying that in the barrier mountain landscape the conductivity is given by the height of the lowest pass, which has to overcome to go across the mountain landscape.

The mathematical condition for obtaining E_{critical} is just Eq. (4.66) with $E_{\text{critical}} = E_a$.

From our analysis it follows that in a disordered material like a ion-conducting glass [9], where the moving ions have to overcome barriers with a broad distribution, the net conductivity will be always of Arrhenius form

$$\sigma(T) = \sigma_0 e^{-E_a/k_B T} \tag{4.67}$$

This result can neither be obtained from averaging the local rates W_{ij} nor from averaging the inverse of them. This corresponds to the well-known result of network

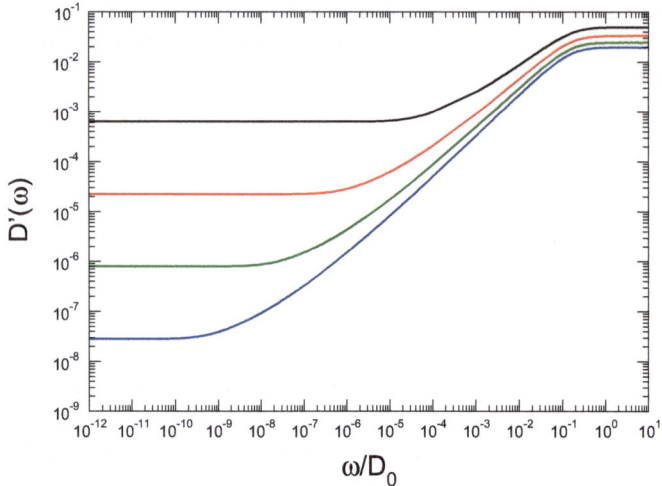

Fig. 4.5 *ac* conductivity for the random-barrier model, calculated in continuum CPA for temperature parameters $\beta^* = 20, 30, 40, 50$ and 60

theory that the net conductance of a complex conductance network can neither be represented by a parallel equivalent circuit nor a series one [7].

A considerable advantage of the CPA as compared to the old effective-medium theories [7] is that one can calculate the *ac* conductivity as well. But now one must decide definitely, which distribution of activation energies one wants to take. The most simple model function is given by

$$P(E) = \text{const} = \frac{1}{E^*} \quad \text{for } E \leq E* \tag{4.68}$$

One easily verifies by changing the integration variable from E to $\epsilon = E/k_B T$, that the resulting CPA equation depends only on the single parameter $\beta^* = E*/k_B T$. In Fig. 4.5 we have plotted the result for the *ac* conductivity $D'(\omega) = Re\{D(s)\}_{s=i\omega+\varepsilon}$ for four values of this temperature parameter. It can be verified easily that for the relative variance of the distribution of local diffusivity holds

$$\frac{1}{\langle D \rangle^2}\langle D^2 \rangle = \frac{1}{2}\beta^* \tag{4.69}$$

So one can say that with decreasing temperature the disorder of the activated diffusivities is increasing.

Frequency-dependent conductivity data in disordered semiconductors and ion conductors look precisely as in Fig. 4.5, as can be verified from the corresponding literature [4, 6, 9].

References

1. Ashcroft, N.W., Mermin, N.: Solid State Physics, Saunders College Publications, Philadelphia, (1976)
2. Born, M., Wolf, E.: Principles of Optics. Pergamon Press, Oxford (1993)
3. Bruggeman, D.: Ann. Phys. **416**(7), 636 (1935)
4. Dyre, J., Schrøder, T.B.: Rev. Mod. Phys. **72**, 873 (2000)
5. Efros, A.I., Shklovskii̇̆, B.I.: Electronic Properties of Doped Semiconductors. Springer, Heidelberg (1984)
6. Jonscher, A.K.: Dielectric Relaxation in Solids. Chelsea Dielectric, London (1983)
7. Kirkpatrick, S.: Rev. Mod. Phys. **45**, 574 (1973)
8. Köhler, S., Ruocco, R., Schirmacher, W.: Phys. Rev. B **88**, 064203 (2013)
9. Wong, J., Angell, C.A.: Glass: Structure by Spectroscopy. Marcel Dekker, New York (1976) [ISBN 0824764684]

References

Chapter 5
Fractals

5.1 Fractal Dimension

In his book "The fractal geometry of nature" [4], which appeared in 1977, the French mathematician Benoit Mandelbrot coined the word *fractal* for geometrical objects, which do not have an inherent length scale, i.e., they are *self similar*. They just look the same at very different length scales. Such object are trees, sponges, termination deltas of rivers, tidal streaming traces, clouds, mountains—and coast lines. Although Mandelbrot made fractals a fashionable subject in physics, self-similar objects have been studied much earlier, e.g., some 150 years ago by people like David Hilbert, Giuseppe Peano or Georg Cantor.

Let us start with coast lines. How long is the coast line of England/Scotland? You may measure it with conventional geodesic wooden sticks to come up with a number of about ten thousand km. (The figure given by www.coastalguide.org is 13,560 km.) However, if you make your measuring device smaller, so that you can follow all small wrinkles you might be able to double the number: The length of the coast line depends on the scale of the measuring device, or, in other words, if you want to draw the coast line its length depends on the thickness or the sharpness of the pencil. In fact a coastline is a typical fractal object: It has similar wrinkles at different length scales.

In discussing the length of the coast line we found that there is some difficulty to identify it as a one-dimensional object, as it has a typical property of an object with dimensionality greater than one: It length depends on something else: For an area this is the width, for the coast line it is the thickness of the pencil. In fact fractals turn out *not to have an integer dimensionality*. Its dimensionality is a *non-integer* number d_f, which is called *fractal dimension*. Let us resume, how in "normal" geometry the dimension is defined: If we multiply the linear size L of a d-dimensional object by a factor b the mass of the object changes by a factor b^d:

$$M(bL) = b^d M(L) \tag{5.1}$$

W. Schirmacher, *Theory of Liquids and Other Disordered Media*, Lecture Notes in Physics 887, DOI 10.1007/978-3-319-06950-0_5,
© Springer International Publishing Switzerland 2015

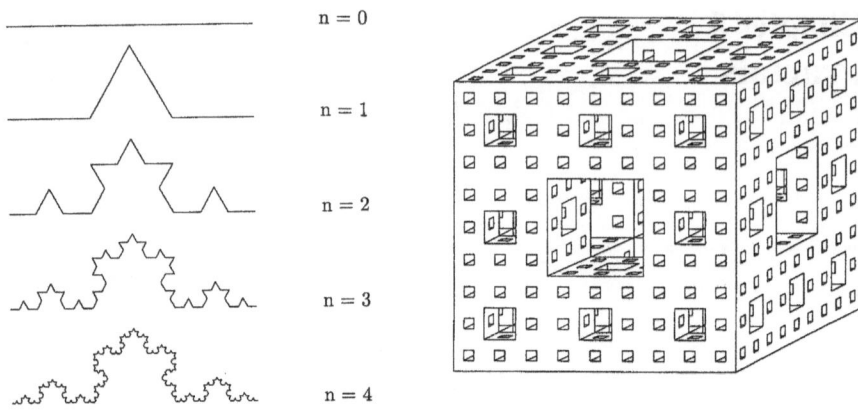

Fig. 5.1 *Left*: Four iterations of the Koch curve. *Right*: Sierpinski sponge

Let us discuss a regular geometric coast line, which is the *Koch curve* depicted in the left part of Fig. 5.1. A straight line is divided into three and the inner part is replaced by the upper part of an equilateral triangle. This procedure is repeated for all four new sides. As the Koch curve is iteratively constructed from lines, i.e., one-dimensional objects its "mass" is just its length. This length steadily increases as the iteration is continued, just as in the coastline example. For the Koch curve we can deduce the value of the fractal dimension: Every time the length is increased by a factor of 3 its length increases by a factor of 4. If we call the length of the Koch curve also M we have

$$M(3L) = 4M(L) \tag{5.2}$$

We want to define the fractal dimension just as in (5.1)

$$M(bL) = b^{d_f} M(L) \tag{5.3}$$

comparing (5.2) with (5.3) we obtain

$$d_f = \ln 4/\ln 3 = 1.26185954\cdots \tag{5.4}$$

One even can generate fractals with dimensions *below* 1. These are point-like objects, called *dusts* by Mandelbrot. The *Cantor set* is iterated by taking just the middle third out of the unity interval, and then this procedure is repeated for every remaining interval. For the remaining dust we have the scaling relation

$$M(3L) = 2M(L) \quad \Leftrightarrow \quad d_f = \ln 2/\ln 3 = 0.630929768\cdots \tag{5.5}$$

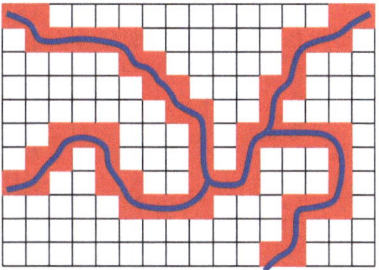

Fig. 5.2 Box-counting determination of the fractal dimension

One can show that from a topological point of view the Cantor set has the Lebesgue measure 0, but its elements are *not* countable, i.e., it can not be mapped onto the set of integers.

A similar procedure can also carried out with a square or a cube. For the latter (*"Sierpinski sponge"*, Fig. 5.1, right part)

$$M(3L) = 20M(L) \quad \Leftrightarrow \quad d_f = \ln 20/\ln 3 = 2.72683311\cdots \quad (5.6)$$

Note that this number is now between 2 and 3.

In cases, in which the scaling law is not obvious one can calculate d_f *empirically* by the so-called *box counting algorithm*. For this we need to define the *imbedding dimension*, which is just called d. The imbedding dimension is the dimension of the space, in which the defining algorithm of the fractal is formulated. So for the Cantor set $d = 1$, for the Koch curve $d = 2$ and for the Sierpinski sponge $d = 3$. For the fractal, for which one wants to determine d_f one sets up a mesh of boxes inside a big hypercube of length L, which are hypercubes of "mass" $(\epsilon L)^d$, where $\epsilon = L/N$ and N is the number of boxes along one edge of the big box (Fig. 5.2).

The box-counting dimension is then defined as

$$d_f = \lim_{\epsilon \to 0} \frac{\ln[M(\epsilon L)/M(L)]}{\ln \epsilon} \quad (5.7)$$

An important fractal, which we encountered in the last sections and which we shall find useful as a simplified model for a polymer coil is a *random walk*. If we interpret the trace of the random walk as a wrinkled one-dimensional object, imbedded into a higher-dimensional space of dimension d, the mass of the object will obviously be proportional to its length $M = Na = t/\tau$, where a is the lattice constant, N is the number of steps leading from one end to the other and t is the time this itinerary would take with time steps τ. On the other hand, the *linear* dimension of the randomly coiled trace is proportional to the square-root of the mean-square distance of the walk

$$L(N) \propto N^{1/2} \quad \Leftrightarrow \quad N(L) \propto L^2, \quad (5.8)$$

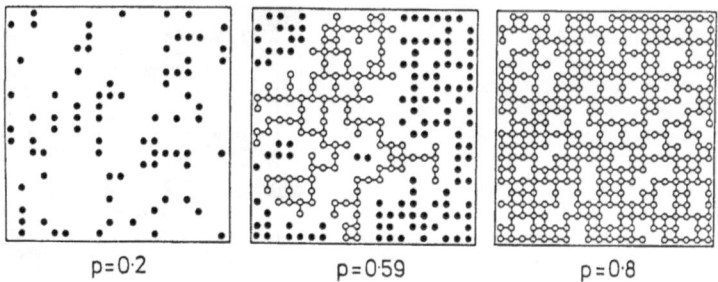

p=0·2   p=0·59   p=0·8

Fig. 5.3 Percolation clusters for three different concentrations p

which means that the fractal dimension of a random walk is $d_f = 2$ in any imbedding dimension d. In contrast to the regular fractals encountered previously in this section the random walk is an example of a *random fractal*, i.e., the self-similarity and scaling law holds only *on the average*. This is true for almost all *real fractals* like mountains, trees and sponges.

One can show that the *static structure factor* has a small-q dependence according to

$$S(q) \propto q^{-d_f} \tag{5.9}$$

This means that we can measure the fractal dimension of a real fractal by X-ray or neutron diffraction.

5.2   Percolation

An important model for a metal-nonmetal transition in a random mixture of a metal with a nonmetal is the *Percolation model*. At the same time it is a "toy model" for a thermodynamic phase transition. It has two versions, namely the *site percolation* model and the *bond percolation* model. In the site percolation model the sites of a d-dimensional lattice is occupied randomly by metal atoms according to the concentration p. If two neighboring sites are occupied they are called *connected*. Connected sites form a *cluster*. If a cluster extends through the system of linear extension L it is called percolation cluster. The critical concentration p_c for the metal-nonmetal transition is defined to be the smallest concentration for which a percolation cluster exists in the limit $L \to \infty$ (Fig. 5.3).

In the bond percolation model *bonds* are randomly distributed on the elementary bonds on the lattice. All sites belonging to a bond are metallic sites and belong to a cluster of connected sites.

The percolation concentrations p_c depend not only on dimensionality but on the type of lattice and whether we have site percolation [5].

Right at $p = p_c$ the percolation cluster forms a fractal. The fractal dimension is *universal* as it depends only on the embedding dimension. For $d = 2$ we have $d_f =$

Table 5.1 Fractal
dimensions d_f of the site
and bond percolation model
[5]

	Site	Bond
$d = 1$		
Square l.	0.6	$\frac{1}{2}$
Triangular l.	$\frac{1}{2}$	0.35
Honeycomb l.	0.7	0.65
$d = 3$		
Simple c.	0.31	0.25
f.c.c	0.20	0.12
b.c.c	0.25	0.18
Simple c.	0.31	0.25
Diamond	0.43	0.39

Table 5.2 Critical exponents β and ν corresponding
to the order parameter $P(p)$ and correlation length
$\xi(p)$. (Bond percolation model [5])

Lattice	β	ν
Quadratic	$\frac{5}{36}$	$\frac{4}{3}$
Simple c.	0.417	0.875

1.9, for $d = 3$ $d_f = 2.55$. The fractal dimensions for site and bond percolation for
some selected lattices are given in Table 5.1

As in thermodynamic phase transitions one can define an *order parameter* $P(p)$,
which is the probability of a site to belong to the percolation cluster. Obviously
$P(p) = 0$ for $p < p_c$ (Tables 5.1 and 5.2). For $p \geq p_c$ we have

$$P(p) \propto (p - p_c)^\beta \tag{5.10}$$

for p near p_c.

For $p \approx 1$ the percolation cluster is obviously not a fractal, as there are only a few
vacancies which do not involve a scaling law. As in the theory of phase transitions
one can define a *correlation length* $\xi(p)$, which has the property that for length
scales $L < \xi$ the percolation cluster looks like a fractal, i.e., $M(L) \propto L^{d_f}$, whereas
for $L > \xi$ $M(L) \propto L^d$ holds. Near p_c we have the critical law

$$\xi(p) \propto (p - p_c)^{-\nu} \tag{5.11}$$

Some critical exponents for bond percolation are shown in Table 5.2.

5.3   Random Walk on a Fractal

One can define a random walk on a fractal just like that on a lattice by defining paths
on the fractal which are allowed for the walker. For the Koch curve the path is just
the "coast line". For the percolation cluster the paths are along the metallic bonds.

It turns out that the mean-square distance walked on a fractal does, in general, *not* increase linearly with time but according to

$$\langle [\mathbf{r}(t) - \mathbf{r}(0)]^2 \rangle \propto t^a \tag{5.12}$$

with $a < 1$. Such a behavior is called *anomalous diffusion*.

5.3.1   Vibrations on a Fractal and Spectral Dimension

Suppose we fix masses m at the nodes of a fractal network, which are connected by springs along the bonds of the network.

Then we have an equation of motion for vibrational elongations $u_n(t)$ (which are assumed to be scalar quantities):

$$\frac{d^2}{dt^2} u_n(t) = -\frac{1}{m} \sum_{\substack{\ell \\ \text{n.N.}}} K_{n\ell}[u_n(t) - u_\ell(t)]. \tag{5.13}$$

For the Fourier transform $u_i(\omega)$ we have

$$-\omega^2 u_n(\omega) = -\frac{1}{m} \sum_{\substack{\ell \\ \text{n.N.}}} K_{n\ell}[u_n(\omega) - u_\ell(\omega)]. \tag{5.14}$$

We can compare this to the master equation for the Fourier transform of the random walk probability $P_i(\omega)$, Eq. (4.43)

$$i\omega P(x_n, \omega) = -\sum_{\substack{\ell \\ \text{n.N.}}} W_{n\ell}[P(x_n, \omega) - P(x_\ell, \omega)] \tag{5.15}$$

Obviously the equations are the same for the random walk and the vibration problem, only the *spectral parameter* for the random walk is $\mathcal{E} = i\omega$, that for the vibrations is $\mathcal{E} = -\omega^2$.

We can therefore expect that the spectral parameter for both problems obeys the same scaling relation, namely

$$\mathcal{E}(L) \propto \frac{1}{t}(L) \propto L^{-\frac{2}{a}} \quad \Rightarrow \quad \omega(L) \propto L^{-\frac{1}{a}} \tag{5.16}$$

where the relation on the right holds for the vibration problem.

We now define a *spectral dimension* d_s for the *vibrational density of states* as follows:

$$\rho(\omega) \propto \omega^{d_s-1} \tag{5.17}$$

For the *integrated* density of states $F(\omega) = \int_0^\omega d\tilde{\omega}\rho(\tilde{\omega})$ we obviously have

$$F(\omega) \propto \omega^{d_s} \tag{5.18}$$

This quantity is just the number of states with frequencies less or equal ω. As the total number of states must be equal to the total number of masses, the number of sites participating with vibrations up to ω must also scale with the mass:

$$F(bL,\omega) = b^{d_f} F(L,\omega) \tag{5.19}$$

On the other hand, we have

$$F(L,\omega) = F(bL, b^{-1/a}\omega) = b^{d_f} F(L, b^{-1/a}\omega) \tag{5.20}$$

Taking $b = A\omega^a$ we obtain

$$F(L,\omega) = A^{d_f}\omega^{d_f a} F(A), \tag{5.21}$$

from which follows

$$d_s = a d_f \tag{5.22}$$

If the random-walk exponent a is less than 1 we have $d_s < d_f$.

5.3.2 The Vibrational Spectrum of Percolation Networks

The spectral dimension d_s is never found to be larger than 2. For percolation in $d = 2, 3$ one has $d_s \approx 1.3$, which led Alexander and Orbach [1] to the conjecture that $d_s = 4/3$ might be an exact relation, but they were rebutted by numerical simulations. However, in many fractal models and realistic fractals there is some evidence that very often $d_s \approx 1.3$. An extensive numerical study of vibrations on percolation clusters in $d = 2$ and $d = 3$ has been performed by Nakayama and Yakubo (see Fig. 5.4). At $p = p_c$ they obtain for both systems a density of states $\rho(\omega) \propto \omega^{d_s-1}$ with $d_f \approx 1.3$.

It is interesting that the *correlation length*, which is finite for $p > p_c$ is reflected in the density of states. For length scales larger than $\xi(p)$ one expects a density of states $\rho(\omega) \propto \omega^{d-1}$ (Debye law). In the homogeneous medium sound waves of a

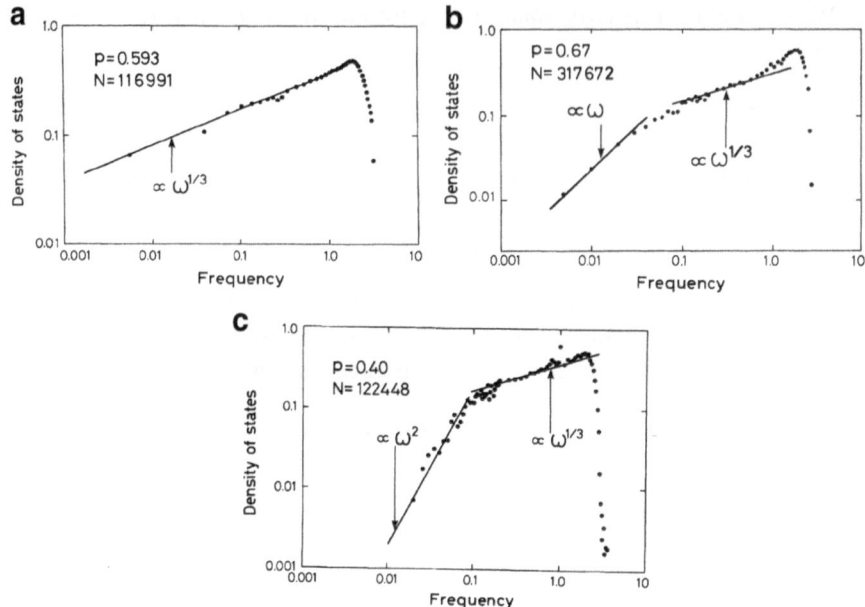

Fig. 5.4 Density of vibrational states for a $d = 2$ percolating network for (**a**) $p = p_c = 0.593$ (right at percolation, $p = pc$) and (**b**) $p = 0.67$ (away from percolation, $p > p_c$). (**c**) Density of vibrational states for a $d = 3$ system, $p = 0.4 > p_c$, from Nakayama et al. [5]

certain sound velocity v are expected to exist, so that the $\rho(\omega) \propto \omega^{d-1}$ law should be valid above $\omega_\xi \approx 2\pi v/\xi$. One can state [1,5]

$$\rho(\omega) \propto \begin{cases} \omega^{d-1} & \omega < \omega_\xi \\ \\ \omega^{d_s-1} & \omega < \omega_\xi \end{cases} \tag{5.23}$$

This is clearly observed in the simulations, both in $d = 2$ and $d = 3$. In $d = 3$ $p_c \approx 0.3$. Alexander and Orbach [1] called such a crossover a phonon-fracton crossover. The fractal vibrational excitations were called fractons.

5.4   The *ac* Conductivity of a Fractal

The function $D(\omega)$ can be related to the *dynamic conductivity* (Appendix C)

$$\sigma(\omega) = i\omega\epsilon(\omega) = \sigma'(\omega) + i\sigma''(\omega). \tag{5.24}$$

$\epsilon(\omega) = \epsilon'(\omega) - i\epsilon''(\omega)$ is the complex permittivity. The real part $\sigma'(\omega)$ is the *ac* conductivity. For $\sigma(\omega)$ we have the *generalized frequency dependent Nernst–Einstein relation*

$$\sigma(\omega) = e^2 \frac{n}{k_B T} D(\omega). \tag{5.25}$$

We have noted that in a fractal the mean-square distance of a random walker does not increase linearly but sublinearly with time with a characteristic exponent a

$$\langle \Delta \mathbf{r}(t)^2 \rangle \propto t^a \tag{5.26}$$

with $a = d_s/d_f$. For the Laplace transform of the mean square distance one obtains

$$\Delta r^2(s) \propto \int_0^\infty e^{-st} dt\, t^a = \frac{1}{s^{a+1}} \Gamma(a+1) \tag{5.27}$$

For the frequency dependent diffusivity we obtain

$$D(\omega) \propto s^{1-a} \tag{5.28}$$

From this we obtain an *ac* conductivity of the form

$$\sigma'(\omega) \propto \omega^{1-a} \tag{5.29}$$

We have now described the *ac* conductivity, due to the polarization phenomena, which is produced by a charged random walker on a fractal. This result holds, however, only if the random walker *can reach all parts of the fractal*, as it is true, for example, in the Sierpinski sponge. However, on a percolating lattice, as pointed out by Gefen et al. [3], there exist isolated clusters, even if the bond or site concentration is above the percolation threshold. If one places a charged random walker on such an isolated cluster, it does not contribute to the *dc* conductivity, because, via the Einstein relation, the *dc* conductivity is related to diffusion across the whole sample. At finite frequencies, however, the motion of charges on the isolated clusters do contribute to the *ac* conductivity. Gefen et al. [3] estimated the influence of the finite clusters as

$$\tilde{\alpha} = \alpha \left(1 + \frac{\beta}{2\nu} \right). \tag{5.30}$$

Fig. 5.5 Log-log representation of the X-ray small-angle scattering curves of porous silicon with three different porosities (volume fraction of "holes"): (a) 55 %, (b) 68 %, (c) 85 %; from Vezin et al. [6]

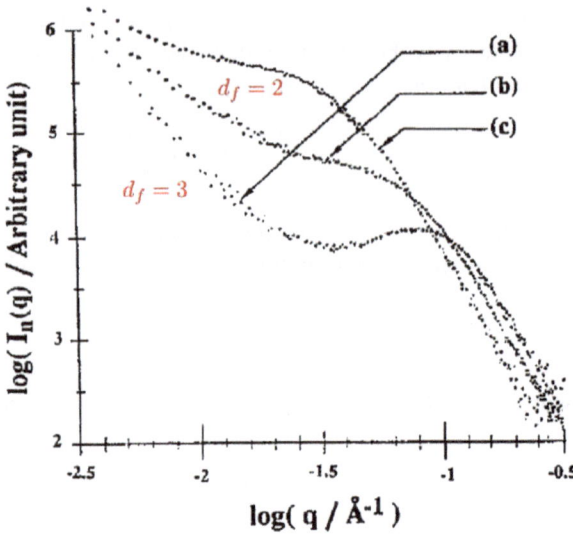

5.5 *ac* Conductivity of Porous Silicon

Porous silicon was investigated extensively in the 1990s of last century, hoping for photovoltaic applicability, as the porous material has a wider band gap than the crystalline one (see [2] for references).

In Fig. 5.5 the structure factor $S(q)$, measured by X-ray diffraction by Vezin et al. [6], is shown, which reveals a self-similar (fractal) structure of the investigated materials. The observed fractal dimensionality depends on the porosity, controlled by the etching procedure.

Let us now look at the *ac* conductivity data of porous silicon, measured by Ben-Chorin et al. [2], as depicted in Fig. 5.6 at different temperatures. It is remarkable (as can be seen from the right figure) that the $\sigma'(\omega)$ can be scaled such that the curves fall on top onto each other. Such a behavior is called *time-temperature superposition property*. It helps to obtain a larger frequency window for $\sigma'(\omega)$.

For small frequencies one observes an *ac* conductivity $\sigma'(\omega) \propto \omega^{1/2}$. Such a behavior corresponds to anomalous diffusion with $a = 1/2$.[1] From the small angle data (Fig. 5.5) we see that in porous silicon one can only speak of "decent" fractal properties for porosities less than $\approx 70\,\%$. For those materials the fractal dimension lies between 2 and 3. According to the *ac* data the spectral dimension should lie between 1 and 2, which is consistent with what one expects.

It is interesting to note that—like in the vibrational spectrum of a fractal—there exists a fractal-non-fractal crossover due to the finite correlation length ξ above this crossover a $\sigma'(\omega) \propto \omega^1$ law ("constant loss ϵ''") is observed:

---

[1] As there are no isolated clusters supposed to be in a porous material, $\tilde{\alpha} = \alpha$.

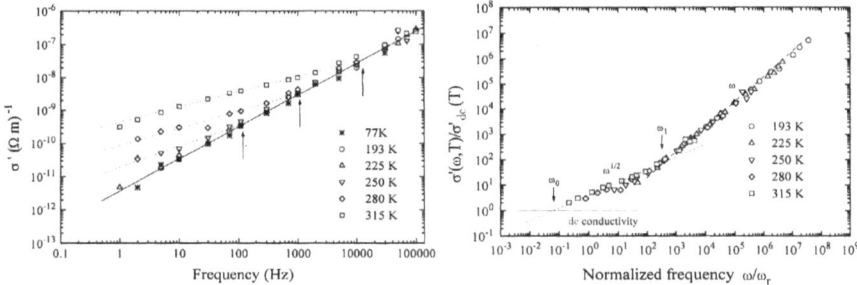

Fig. 5.6 *ac* conductivity data of porous silicon at different temperatures as measured (*left*); scaled (*right*) from Ben-Chorin et al. [2]

$$\sigma'(\omega) \propto \begin{cases} \omega^{1-a} & \omega < \omega_\sigma \\ \omega^1 & \omega < \omega_\sigma \end{cases} \tag{5.31}$$

The *ac* behavior in the "non-fractal" regime can be explained in terms of activated hopping between defect pairs with distributed activation energies. The crossover frequency ω_σ can be related to the time in which the anomalous $\langle \Delta r(t) \rangle$ increases from a microscopic length to the correlation length. From the small-angle data we estimate this length difference to be roughly one order of magnitude. Using $\langle \Delta R(t)^2 \rangle \propto t^{1/2}$ we obtain a "fractal" time or frequency window of four orders of magnitude. This corresponds nicely with the experimental findings.

5.6 The Fractal Dimension of a Self-avoiding Walk

In Sect. 5.1 we mentioned that the *random walk* if studied as a function of the steps N is, in fact a fractal with $d_f = 2$ in any imbedding dimension d. Such an object might be a model for a polymer, if—and this is an important if—it were not for the fact that a polymer cannot intersect itself, i.e., the chain cannot occupy more than once the same portion of space. A random walk which never uses the same site is has already used is called a *self-avoiding random walk*. In a real polymer the excluded-volume property is, of course, due to a repulsive potential $\phi(|\mathbf{r} - \mathbf{r}'|)$ between two monomer units at locations $\mathbf{r}$ and $\mathbf{r}'$. Flory has calculated the relation between the length N and the extension R of a self-avoiding random chain by a thermodynamic argument. The probability density for the end of the chain of length N to be a distance R from the other end (origin) is, using (4.25) with $2Dt = a^2 t/\tau = a^2 N$:

$$P(R, N) = [2\pi a^2 N]^{-3/2} e^{-R^2/2a^2 N} \tag{5.32}$$

The number of random walks with N steps leading from the origin to any point inside a sphere of radius R^* is then given by

$$\zeta(R^*, N) = 4\pi \int_0^{R^*} R^2 P(R, N)$$ (5.33)

From this we can calculate the number of random walks having exactly the distance R from the origin as

$$Z(R, N) = \frac{d\zeta}{dR^*}\bigg|_{R^*=R} = 4\pi R^2 P(R, N)$$ (5.34)

The corresponding entropy is

$$S(R, N) = k_B \ln[Z(R, N)] = -k_B \frac{R^2}{2a^2 N} + k_B \ln[4\pi R^2] - (3/2)k_B \ln[2\pi a^2 N]$$
(5.35)

We now estimate the mean repulsive energy as follows:

$$E = \rho_0^2 \int_V d^3\mathbf{r} \int_V d^3\mathbf{r'} g(|\mathbf{r} - \mathbf{r'}|)\phi(|\mathbf{r} - \mathbf{r'}|)$$ (5.36)

Here $V = R^3$, $g(r)$ is the radial pair distribution function of the monomers and $\rho_0 = N/V = NR^{-3}$ is their density. As $g(r)$ is 0 for $|\mathbf{r} - \mathbf{r'}| < d$ (where $d \approx a$ is the distance of nearest approach) and $\phi(r)$ is supposed to drop quickly to 0 beyond d we can make the approximation

$$g(r)\phi(r) \approx \epsilon\delta(r - d),$$ (5.37)

where ϵ has the dimension of an energy times a volume. We obtain

$$E = \rho_0^2 V\epsilon = \epsilon N^2/R^3$$ (5.38)

We can now write down the free energy

$$F = E - TS$$
$$= \epsilon N^2/R^3 + k_B T \ln[Z(R, N)]$$
$$= \epsilon N^2/R^3 + k_B T \frac{R^2}{2a^2 N} - k_B T \ln[4\pi R^2] - (3/2)k_B \ln[2\pi a^2 N] \quad (5.39)$$

We now seek the equilibrium value of R for a self-avoiding random walk of N steps, which, is obtained by that value of R which minimizes F, i.e.,

Fig. 5.7 DLA cluster from
the original paper of Witten
and Sander [7]

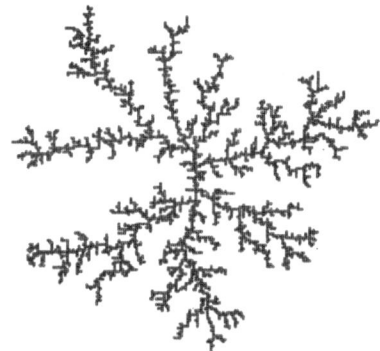

$$0 = \frac{\partial F(R, N)}{\partial R} = -3\frac{\epsilon N^2}{R^4} + \frac{k_B T R}{a^2 N} - \frac{2k_B T}{R}, \qquad (5.40)$$

from which follows

$$\frac{R^2}{Na^2} = 3\frac{N^2\epsilon}{R^3 k_B T} + 2 \qquad (5.41)$$

In the limit of large N and R (and for temperatures equal or smaller than $\epsilon N^2/R^3$) the constant term 2 is negligible, and we obtain

$$N(R) = \left[\frac{k_B T}{3\epsilon a^2}\right]^{1/3} R^{5/3} \qquad (5.42)$$

5.7 Diffusion-Limited Aggregation

An interesting "random" fractal object is obtained in the following way: Consider a large two-dimensional square lattice with an approximately spherical boundary on which random walkers are started with equal probability. If these walkers reach the boundary, a further walker is started. If one of the walkers reaches the origin, the site from which it arrived at the origin is attached to the origin as a second site of a *beginning cluster*. If another walker reaches one of the cluster sites the site from which it arrives is added to the cluster. The cluster, which grows in this way looks like a tree and is a fractal object with fractal dimension $d_F \approx 1.7$ in $d = 2$ and $d_F \approx 2.5$ in $d = 3$ (Fig. 5.7).

The "crumpled" form of the cluster arises because there exists an intrinsic *growth instability* at the surface of the cluster: The local growth rate is much larger if the local surface is *curved*, i.e., if a site on the cluster can be reached by more than one free site. This situation is common to *dendritic growth* of a crystal from a liquid with temperature below the melting temperature or from a solution with a slowly diffusing solvent.

References

1. Alexander, S., Orbach, R.: J. Phys. (Paris) **43**, L685 (1982)
2. Ben-Chorin, M., Müller, F., Koch, F., Schirmacher, W.: Phys. Rev. B **51**, 2199 (1995)
3. Gefen, Y., Aharony, A., Alexander, S.: Phys. Rev. Lett. **50**, 77 (1983)
4. Mandelbrot, B.: The Fractal Geometry of Nature. Freeman, New York (1977)
5. Nakayama, T., Yakubo, K., Orbach, R.L.: Rev. Mod. Phys. **66**, 381 (1994)
6. Vezin, V., Goudeau, P., Naudon, A., Halimaoui, A., Bomchil, G.: Appl. Phys. Lett. **60**, 2625 (1992)
7. Witten, T.A., Sander, L.M.: Phys. Rev. Lett. **47**, 1400 (1981)

Chapter 6
Structure of Polymers

6.1 Single Ideal Polymer Chain

On a microscopic level a polymer chain mainly consists of a *backbone* of beads to which some *side groups* are attached. In addition there can be *branching* and *cross-linking*.

There is usually a very limited range of possible *bond angles* between the beads. The angle θ between the beads is usually fixed, but the *azimuthal* or *dihedral* angle ϕ can have different values, e.g., $\phi = 0°$ (*cis*), $\phi = 60°$ (*gauche*), or $\phi = 180°$ (*cis*), where the latter is usually the most stable position (Fig. 6.1). Between the bond angle positions there are energetic barriers $\Delta\epsilon$, which the chain has to overcome for changing the angle. If the temperature is much higher than these energy barriers, the chain will acquire some freedom to form a curvature, i.e., the directions of the beads will start to fluctuate statistically. Let us denote by $\mathbf{a}_n$ the vector pointing from one chain connection to another. Let us assume that they have all the same length a. Then we can be interested in the following correlation function

$$C(v) = \langle \mathbf{a}_{n+v} \cdot \mathbf{a}_n \rangle \tag{6.1}$$

This function will decay exponentially with increasing v with a characteristic decay constant $v_0 = \ell_0/a$, where ℓ_0 is the decay length. For length scales larger than ℓ_0 the chain will behave as an *ideal polymer chain* i.e., like a *random walk* if we for a moment disregard the volume exclusion condition.

In particular, we can calculate the *radius of gyration* R_0, which is the square-root of the mean square end-to-end distance:

$$R_0^2 = \left\langle \left[\sum_{n=1}^{N} \mathbf{a}_n \right]^2 \right\rangle = Na^2 + \sum_{n=1}^{N} \sum_{v=1}^{v_{max}} [C(v) + C(-v)] = Na^2 \tag{6.2}$$

W. Schirmacher, *Theory of Liquids and Other Disordered Media*, Lecture Notes in Physics 887, DOI 10.1007/978-3-319-06950-0_6,
© Springer International Publishing Switzerland 2015

Fig. 6.1 Three segments of a
polymer chain with bond
angles θ and azimuthal angle
ϕ by which the bond can be
directed into the *trans* and the
gauche direction

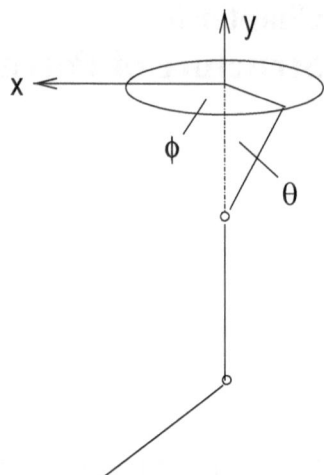

The second equality holds, because the correlations have all values between a^2
and $-a^2$ with equal probability, i.e they average to zero.

So we can state that a polymer chain at a temperature above the typical energy
barriers for bond-angle torsion behaves like an ideal chain as modelled by a random
walk.

As starting point for a thermodynamic treatment of a polymer chain we can
therefore take the number of possible random walks with N steps and end-to-end
distance R as calculated in the previous subsection:

$$Z(R, N) = 4\pi R^2 [2\pi R_0^2]^{-3/2} e^{-R^2/2R_0^2} \tag{6.3}$$

with corresponding entropy

$$S_{id}(R, N) = k_B \ln[Z(R, N)] = -k_B \frac{R^2}{2R_0^2} + k_B \ln[4\pi R^2] - (3/2)k_B \ln[2\pi R_0^2] \tag{6.4}$$

and free energy

$$F_{id} = -T S_{id} = k_B T \frac{R^2}{2R_0^2} - k_B T \ln[4\pi R^2] - (3/2)k_B \ln[2\pi R_0^2] = F_0 + k_B T \frac{R^2}{2R_0^2} \tag{6.5}$$

From this we can calculate the distance x the chain will be elongated if an
external force f in x direction is applied. The corresponding potential is $\phi_f = -fx$
so that the free energy is

$$F_f = -fx + F_0 + k_B T \frac{x^2 + y^2 + z^2}{2R_0^2} \tag{6.6}$$

Minimizing F_f with respect to x yields

$$x = f \frac{R_0^2}{k_B T} = f \frac{Na^2}{k_B T} \tag{6.7}$$

This equation can also be obtained by a *scaling argument*:

(i) As we suppose that the force f acts on the entire polymer chain we expect that the elongation x is a linear function of N.
(ii) The only relevant variables x may depend on are R_0, f and T.

As R_0 is the only relevant length scale we can write

$$x = A R_0 \left(\frac{f R_0}{k_B T} \right)^\alpha \tag{6.8}$$

The exponent α is fixed by the requirement (i) which states that x should be proportional to R_0^2, i.e., $\alpha = 1$. The proportionality constant A cannot be fixed by the scaling argument.

Next we are interested in the behavior of a polymer chain which is confined between two parallel plates of distance D. We assume that the inner walls of the plates repel the polymer so that there are no adsorption effects.

Now, as our model for an ideal chain is a random walk as introduced in the last section, the walk in three dimensions is a superposition of three independent walks in x-, y- and z direction. If the confining plates are parallel to the x-y plane we expect that the mean extension r_0 of the polymer in the x-y plane is given by

$$r_0 = R_0 \tag{6.9}$$

By the same token the extension x_0 of an ideal polymer chain inside a *tube of diameter D* is expected to be

$$x_0 = R_0. \tag{6.10}$$

We now try to estimate the free energy required to squeeze the polymer between the two plates of diameter D. We use a scaling argument similar to that for the force and start by stating that the only relevant quantity will be the *entropy change ΔS*:

(i) ΔS should be a linear function of N.
(ii) The only relevant variables are R_0 and D, so

$$\Delta S = -A \left(\frac{R_0}{D} \right)^\beta \tag{6.11}$$

Again, from requirement (i) we obtain the exponent, namely $\beta = 2$, so that the free energy of squeezing is

$$\Delta F = Ak_BT \left(\frac{R_0}{D}\right)^2 \tag{6.12}$$

Let us now consider a *single* plate and consider the possibility that an ideal polymer chain may be *adsorbed* to its surface by means of a *contact energy* ϵ which is gained if a monomer unit is attached to the plate. We assume that by this action a confinement of thickness D occurs so that a fraction of a/D monomers have the chance to contact the plate. The free energy for this situation is

$$F = -N\frac{a}{D}\epsilon + k_BT\frac{R_0^2}{D^2}, \tag{6.13}$$

where the second term is again the free energy of squeezing. Minimizing this expression with respect to D we obtain (with $A = 1$)

$$D = 2k_BT\frac{R_0^2}{Na\epsilon} = 2k_BT\frac{a}{\epsilon}, \tag{6.14}$$

from which follows

$$F = -\frac{1}{4}N\frac{\epsilon^2}{k_BT} \tag{6.15}$$

We see that, as to be expected, the free energy of adsorption *decreases*, and the layer thickness *increases* with increasing temperature.

Another quantity of interest is the *radial pair correlation function of monomers* $\tilde{g}(r)$ which is just the counter part of this quantity in simple liquids. In simple liquids $g(r)$ was defined in such a way that for large r the integral

$$\rho_0 4\pi \int_0^{R_{max}} dr\; r^2 g(r) = \frac{4}{3}R_{max}^3\frac{N}{V} = N \tag{6.16}$$

where $\rho_0 = N/V$ is the number density of molecules. Here we identify ρ_0 with the number density of monomers and include the factor ρ_0 in the definition of the correlation function:

$$\tilde{g}(r) = \rho_0 g(r) \tag{6.17}$$

so that we have

$$4\pi \int_0^\infty dr\; r^2 \tilde{g}(r) = N \tag{6.18}$$

We now try to figure out a scaling form of $\tilde{g}(r)$. As this function has the dimension Length^{-3} we make the ansatz

$$\tilde{g}(r) = A\frac{N}{R_0^3} f(\frac{r}{R_0}) \tag{6.19}$$

We now realize that inside a sphere of radius r^* we must have

$$4\pi \int_0^{r^*} dr \; r^2 \tilde{g}(r) = n , \tag{6.20}$$

where n is the number of monomers inside the sphere of radius r^*. According to the random walk rule of the ideal chain we must have

$$(r^*)^2 \propto na^2 , \tag{6.21}$$

which can only be reconciled with (6.20) if the function $f(x)$ is proportional to x^{-1}. We therefore obtain

$$\tilde{g}(r) = A\frac{N}{R_0^3}\frac{R_0}{r} = A\frac{1}{ra^2} . \tag{6.22}$$

This is the so-called *Debye correlation function.* It has the Fourier transform

$$\tilde{S}(k) = 4\pi A\frac{1}{q^2} . \tag{6.23}$$

6.2 Swollen Polymer Chains

We now return to a discussion of more realistic polymer chains in taking the *excluded-volume interaction* into account (section 5.6). In terms of the repulsive excluded-volume energy ϵ the free energy is given by

$$F = F_{\mathrm{id}} + \frac{\epsilon N^2}{R^3} F = F_0 + k_B T\frac{R^2}{2Na^2} + \frac{\epsilon N^2}{R^3} \tag{6.24}$$

which was minimalized to obtain the Flory law

$$R_F(N) = \left[\frac{k_B T}{3\epsilon a^2}\right]^{1/5} N^{3/5} , \tag{6.25}$$

where the index F denotes "Flory". This yields a free energy of the form

$$F = F_0 + \frac{1}{2}k_B T\frac{R_F^2}{2Na^2}(1 + 3) = 2k_B T\frac{R_F^2}{R_0^2} \tag{6.26}$$

The exponent $3/5 = 0.6$ is considerably larger than the free-chain exponent 0.5. One therefore speaks of chains which are "swollen" through the excluded-volume interaction.

We are now going to repeat the scaling calculations of the last section for the case of "real" or "swollen" polymer chains. The *scaling argument* for the external force f now proceeds as follows: We write the elongation $x(f)$ as

$$x(f) = AR_F\varphi\left(\frac{fR_F}{k_BT}\right) \tag{6.27}$$

As we expect that $x \propto f$ for small f the universal function $\varphi(x)$ must start linearly. On the other hand, for large x we expect x to be proportional to N as the force is assumed to act on all monomer units. So $\varphi(x) \propto x^\gamma$ and we must have $1 = \frac{3}{5}(1 + \gamma)$ from which follows $\gamma = \frac{5}{3} - 1 = \frac{2}{3}$. We obtain

$$x(f) = \begin{cases} Af\frac{R_F^2}{k_BT} & f < k_BT/R_F \\ AR_F\left(\frac{fR_F}{k_BT}\right)^{2/3} & f > k_BT/R_F \end{cases} \tag{6.28}$$

We see that in the limit of small f the force to pull the swollen coil apart must be larger than that to pull an idea coil apart. For $f > k_BT/R_F$ the "swelling interaction" is negligible and we have a cross-over to the ideal behavior (which we, of course, put into the scaling argument on the first place!). We now turn to the size of a "sausage" inside a tube of diameter D. We write

$$x_0 = R_F\Phi(R_F/D) \tag{6.29}$$

For $D \to \infty$ we must have $\Phi(0) = 1$. For large arguments of the scaling function we obtain a one-dimensional object so that x_0 must be proportional to $N \propto R_F^{5/3} \Rightarrow \Phi(x) \propto x^{2/3}$ for $x \to \infty$. In this limit we obtain

$$x_0 \propto \frac{N}{D^{2/3}}. \tag{6.30}$$

The size of a "real sausage" (fixed volume V) scales, of course, as $x_0 \propto D^{-1}$. We now estimate the entropy of a swollen polymer between two plates of distance D. Again we postulate that S should be a universal function of R_F/D and proportional to N. Therefore it must scale as

$$S = -A\left[\frac{R_F}{D}\right]^{5/3} \tag{6.31}$$

so that

$$F = Ak_BT \left[\frac{R_F}{D}\right]^{5/3}. \tag{6.32}$$

From this we write down the free energy of an adsorbed polymer pankake as

$$F = -N\frac{a}{D}\epsilon + Ak_BT \left[\frac{R_F}{D}\right]^{5/3} \tag{6.33}$$

Minimization of this expression with respect to D leads to a thickness of the adsorbed layer which scales as

$$D \propto \left[\frac{k_BT}{\epsilon}\right]^{3/2} \tag{6.34}$$

We see that all the scaling laws now have changed due to the fact that the fractal dimension of the self-avoiding walk is $d_F = 5/3$ instead of $d_F = 2$ in the case of the ideal random walk.

The argument that led to the scaling law (6.22) of the *pair correlation function* now leads to the general scaling law for any fractal

$$g(r) \propto r^{d_F - d} \propto r^{-4/3} \tag{6.35}$$

and

$$S(k) \propto k^{-d_F} \propto k^{-5/3}. \tag{6.36}$$

6.3 Polymer Melts

The arguments to the last sections related to a *single polymer chain*. Such single chains (or rather coils) appear in *dilute solutions* of polymers in good solvents (i.e., solvents with small or negative interaction parameters). On the other hand in *concentrated solution* and *melts* it turns out that the self-avoiding interaction is suppressed, so that their scaling behavior is almost that of an ideal random walk. This fact is referred to as "Flory's theorem" and can be made plausible by means of the following argument.

Let us consider a spaghetti-like arrangement of many different polymer chains and consider one particular of these chain, the "tagged chain" or "tagged coil". We consider the repulsive potential energy U experienced by the tagged coil. This energy must be proportional to the concentration $c_{\text{tagged}}(\mathbf{r})$ of monomers of the tagged coil. This concentration is maximal near the center of mass of the tagged coil. Therefore a force is effective towards the outward direction. This force leads to the swelling of the coil in dilute solutions. Now we are in a situation that everywhere the total concentration c_{tot} of monomers of *any* coil is essentially constant, so that the concentration $c_{other} = c_{tot} - c_{tagged}$ of the *other* coils becomes depleted in the spatial region of the tagged coil. This leads to a force *inwards* into the direction of the center of mass of the tagged coil which exactly cancels the force considered previously. In simple words, the tagged coil cannot swell because of the presence of the other ones which claim the same right.

6.4   Polymer Solutions in Good Solvents

In the last section we presented an argument according to which *melts* of polymers behave like ideal random-walk chains, whereas dilute solutions of polymers are swollen, i.e., they scale like self-avoiding walks. It is reasonable that *concentrated solutions* of polymers, i.e., a large amount of polymer solute together with a small amount of solvent will also essentially behave ideally, because the "screening argument" will still hold. However, going from the dilute to the concentrated regime there will be a volume fraction ϕ^* which marks the crossover from the dilute to the concentrated regime. This crossover volume fraction can, of course, also be estimated by a scaling argument.

Let us assume the ideal coils occupy a sphere of radius R_0, the swollen coils a sphere of radius R_F If the packing fraction

$$\eta_F = \frac{\pi}{6} \rho_p R_F^3 \tag{6.37}$$

reaches a value near 0.5 the concentrated limit is reached. $\rho_p = N_p/V$ is the number density of polymers. Because $\phi = a^3 \rho_p N$ we have for the crossover volume fraction

$$1 \approx \rho_p R_F^3 = \frac{\phi^*}{a^3 N} a^3 N^{9/5} = \phi^* N^{4/5} \quad \Rightarrow \quad \phi^* = N^{-4/5} \tag{6.38}$$

The dilute limit in which the polymers are swollen is obviously a *very* dilute regime. The regime in which $\phi^* < \phi \ll 1$ holds is called *semi-dilute* regime. In the dilute limit the Flory–Huggins free Enthalpy $\Delta g = \Delta G / k_B T$, which we now re-formulate as a free energy $\Delta f = \Delta F / k_B T$ can be expanded

$$\Delta f = \phi(1 - \phi)\chi + \frac{\phi}{N} \ln \phi + (1 - \phi)\ln(1 - \phi) \approx \frac{\phi}{N} \ln \phi + (\chi - 1)\phi + \frac{1}{2}(1 - 2\chi)\phi^2 \tag{6.39}$$

It is revealing to relate this to an important physical quantity if one deals with dilute solutions. This is the *osmotic pressure*. It is the pressure due to the dissolved molecules. One usually performs a *Gedankenexperiment* with a semipermeable membrane which is impermeable for the solute but permeable for the solvent. The construction is to add solvent together with the solution volume but leave the number N_p of polymer molecules fixed. The osmotic pressure is defined as the derivative of the free energy in the total volume $\Delta F/\phi$ with respect to the inverse volume fraction:

$$a^3 \Pi = -\frac{\partial \Delta F/\phi}{\partial 1/\phi} = \phi^2 \frac{\partial \Delta F/\phi}{\partial \phi} \tag{6.40}$$

which becomes

$$\frac{a^3}{k_B T} \Pi = \frac{1}{N}\phi + \ln\left(\frac{1}{1-\phi}\right) - \frac{\partial \Delta F/\phi}{\partial 1/\phi} - \phi - \chi\phi^2 \approx \frac{1}{N}\phi + \frac{1}{2}(1-2\chi)\phi^2 \tag{6.41}$$

As $\phi = a^3 N N_p / V$ the first term is the osmotic version of the ideal gas equation of state. In the theory of interacting gases the quadratic term in the density expansion (*virial expansion*) is called the *second virial coefficient* and gives the correction to the ideal gas equation of states due to interactions. However, it turns out that the Flory–Huggins form of the second virial coefficient is *not* correct.

In the *dilute* limit $\phi < \phi^*$ the polymers are swollen and do not want to interpenetrate each other. This leads in the case of an ideal (*"athermal"*) solution with $\chi = 0$ to an equation of state of the form

$$\frac{a^3 \Pi}{k_B T} = \frac{\phi}{N} + \left(\frac{\phi}{N}\right)^2 \left(\frac{R_F}{a}\right)^3 = \frac{\phi}{N}\left[1 + \frac{\phi}{N}\left(\frac{R_F}{a}\right)^3\right] \tag{6.42}$$

As the second virial coefficient is obviously proportional to $N^{-1/5}$ it is strongly reduced in comparison with that of the mean-field expression (6.41). We now look for an extension of the equation of state of the athermal polymer solution into the *semi-dilute* regime $\phi > \phi^*$. Inspired by (6.42) we write

$$\frac{a^3 \Pi}{k_B T} = \frac{\phi}{N} f_s \left(\frac{\phi R_F^3}{Na^3}\right) \tag{6.43}$$

In the very dilute limit $\phi < \phi^*$, according to (6.42) the universal function f_s behaves as $f_s(x) = 1 + x$. As in the semi-dilute regime $\phi > \phi$ the material behaves as a "monomer soup" of volume fraction ϕ, all dependence on the degree of polymerization N should drop out of the equation of state. Writing $f_s(x) \propto x^\delta$ we obtain

$$\frac{a^3 \Pi}{k_B T} \propto \phi^{\delta+1} N^{[\delta(9/5-1)-1]} \tag{6.44}$$

If the exponent of N is to be 0 we must have $\delta = 5/4$ and the volume fraction dependence of the osmotic pressure in the semi-dilute regime becomes

$$\frac{a^3 \Pi}{k_B T} \propto \phi^{9/4} . \tag{6.45}$$

In semi-dilute athermal polymer solutions one can introduce a *correlation length* ξ, just as we did in the case of *real fractals*, i.e., fractals which exhibit their self-similar scaling law only below a length scale ξ:

$$M(L) \propto \begin{cases} L^{d_f} & L < \xi \\ \\ L^d & L > \xi \end{cases} \tag{6.46}$$

As for $\phi > \phi^*$ the coils (which for $\phi \geq \phi^*$ would have radius R_F) are *overlapping* we expect that there exists a length scale ξ, for which

$$R(N) \propto \begin{cases} N^{3/5} & L < \xi \\ \\ N^{1/2} & L > \xi \end{cases} \tag{6.47}$$

ξ can be measured by small-angle neutron or X-ray scattering (see below). We now construct a scaling law for this correlation length. For $\phi \approx \phi^*$ where the single polymer coils just touch, but not yet interpenetrate each other, ξ should be of the order of the Flory length R_F. We write

$$\xi(\phi) = R_F \left(\frac{\phi}{\phi^*} \right)^{\epsilon} \tag{6.48}$$

Again we require that in the semi-dilute regime the correlation length should be independent of N. Therefore the exponent ϵ should be such that the exponent of $R_F \propto N^{3/5}$ and the exponent of $(1/\phi^*)^{\epsilon} \propto N^{\frac{4}{5}\epsilon}$ cancel, which leads to $\epsilon = -3/4$, and we have in the semi-dilute regime

$$\xi(\phi) \propto \phi^{-3/4} . \tag{6.49}$$

We see that the correlation length decreases rapidly, albeit with a power law, from R_F to 0 in the more and more concentrated regime.

It is interesting to note that the scaling law for the osmotic pressure (6.43), (6.44), and (6.45) can be taken together by writing

$$\frac{a^3 \Pi}{k_B T} = \left(\frac{a}{\xi} \right)^3 \tag{6.50}$$

The pair distribution function of monomers for separations $r < \xi$ in semi-dilute solutions must decay according to the fractal law $g(r) \propto r^{d_f - d} = r^{-4/3}$, whereas for $r > \xi$ it should decay according to a Debye law $g(r) \propto r^{-1}$. The corresponding

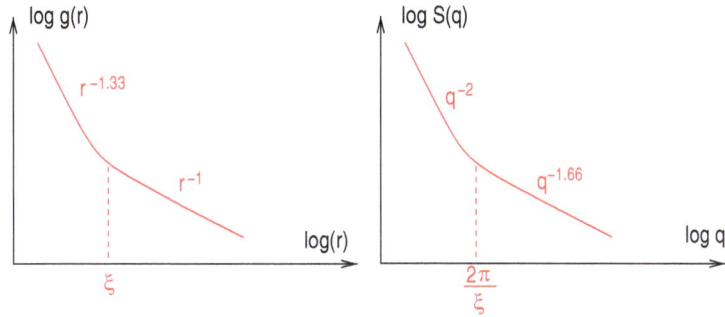

Fig. 6.2 *Left*: radial pair distribution $g(r)$ of monomer units for a semi-dilute polymer solution. *Right*: the corresponding static structure factor $S(q)$

small-angle static structure factor should behave as q^{-2} for $q < 2\pi/\xi$ and as $q^{-d_f} = q^{-5/3}$ for $q > 2\pi/\xi$ (see Fig. 6.2).

6.5   Poor Solvents and Segregation

We now return to the discussion of the segregation properties of the Flory–Huggins model

$$\frac{\Delta f}{k_B T} = \frac{1}{N}\phi \ln \phi + (1-\phi)\ln(1-\phi) + \phi(1-\phi)\chi. \tag{6.51}$$

The spinodal curve, i.e., the χ which marks the borderline of stability is given by

$$0 = \frac{\partial^2}{\partial\phi^2}\frac{\Delta f}{k_B T} = -2\chi_{\text{spinodal}} + \frac{1}{N\phi} + \frac{1}{1-\phi} \tag{6.52}$$

Instead of the spinodal and the coexistence temperatures T_{spinodal} and T_{coex} (double-tangent construction) we have plotted in the picture (Fig. 6.3) its inverse, i.e., $2\chi_{\text{spinodal}} = \omega/k_B T_{\text{spinodal}}$, $2\chi_{\text{coex}} = 2\omega/k_B T_{\text{coex}}$. We see that in the segregation regime the dilute phase has a volume fraction below ϕ^*, indicating that the dilute phase consists of swollen coils, whereas the concentrated phase does not.

We would like to emphasize here that the swelling is an excluded-volume effect (like the hard-core repulsion of the simple liquid constituents), whereas the phase separation is an effect of the *Van-der-Waals attraction*.

Fig. 6.3 Inverse spinodal $\chi_c(\phi)$ for $N = 10000$ together with the coexistence χ line for the Flory–Huggins mean-field model. The *dashed vertical line* indicates the borderline $\phi^* \propto N^{-4/5}$ to the regime of swollen chains, the *dashed horizontal line* is the Θ line $2\chi_\Theta = 2\omega/k_B\Theta = 1$

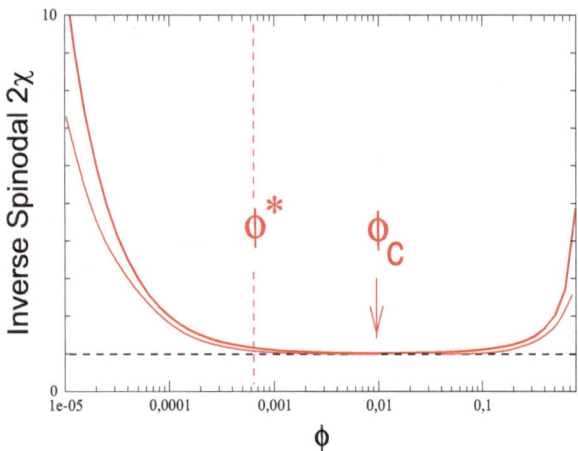

In fact, if we write

$$U_{ij} = -A\alpha_i\alpha_j \tag{6.53}$$

where A is a proportionality constant and the α_i are the polarizability of the solute monomer units ($i = 1$) and the solvent molecules ($i = 2$) we obtain

$$\omega = k_B T \chi = \frac{A}{2}(\alpha_1 - \alpha_2)^2 . \tag{6.54}$$

So that we conclude that good solvents have the same polarisability as the solute, bad solvents have a different one.

6.6   Polymer Mixtures

Of course the Flory–Huggins model can be generalized to the case of a liquid mixture of two polymer species with chain lengths N_A and N_B. One can consider both the situation of two polymer species dissolved in a good solvent as well as a binary polymer melt.

In both cases the free energy of mixing can be written, according to the Flory–Huggins model, as

$$\frac{\Delta f}{k_B T} = \frac{1}{N_A}\phi \ln\phi + \frac{1}{N_B}(1 - \phi) \ln(1 - \phi) + \phi(1 - \phi)\chi , \tag{6.55}$$

where ϕ is the volume fraction of the A species, $1 - \phi$ that of the B species. In the case of a solution of the two species in a solvent ϕ and $1 - \phi$ are the fractions occupied by monomer units relative to the total volume occupied by polymer material. The stability function is given by ($c = c_A = \phi_A/N_A$)

$$f_{cc} = S_{cc}(q = 0) = \frac{\partial^2}{\partial c^2} \frac{\Delta f}{k_B T} = \frac{1}{N_A \phi} + \frac{1}{N_B(1 - \phi)} - 2\chi . \tag{6.56}$$

The spinodal is given by

$$2\chi_{\text{spinodal}} = \frac{1}{N_A \phi} + \frac{1}{N_B(1 - \phi)} \tag{6.57}$$

from which we obtain a critical concentration of

$$\phi_c = \frac{N_B^{1/2}}{N_A^{1/2} + N_B^{1/2}} \tag{6.58}$$

The critical χ is given by

$$\chi_c = \omega/k_B T_c = \frac{1}{2}\left(\frac{1}{N_A^{1/2}} + \frac{1}{N_B^{1/2}}\right)^2 \tag{6.59}$$

For $N_A = N_B = N$ the model is very similar to the conformal solution model. In this case we obtain a critical χ parameter of

$$\chi_c = \frac{2}{N} \tag{6.60}$$

Obviously one can work with a conformal solution model of $\chi_{\text{eff}} = (N/4)\chi$.

Generalizing the random phase approximation which led to the Flory–Huggins theory we can write

$$S_{cc}(q) = \frac{S_0(q)}{1 - 2\chi S_0(q)} \quad \Leftrightarrow \quad \frac{1}{S_{cc}(q)} = \frac{1}{S_0(q)} - 2\chi \tag{6.61}$$

For $S_0(q)$ we can take a Debye law $1/S_0(q) \propto q^2$ from which we get

$$S_{cc}(q) \propto \frac{1}{q^2 + \xi_c^{-2}} \tag{6.62}$$

where the correlation length ξ_c describes the *critical fluctuations* due to the vicinity of the demixing phase transition and is given by

$$\xi_c \propto (\chi_c - \chi)^{-1/2} \propto (T - T_c)^{-1/2} . \tag{6.63}$$

Fig. 6.4 Phase diagram for
equilibrium
diblock-copolymer phases as
a function of the χ parameter
and the volume fraction ϕ_A

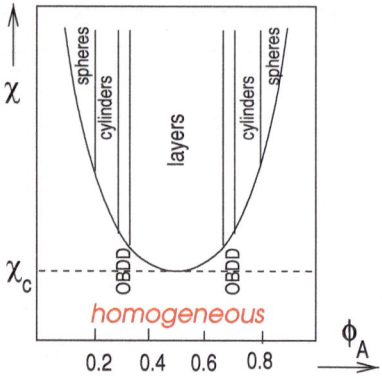

The critical exponent $\nu = 1/2$ is the mean-field exponent. In the vicinity of
the critical point it can be different from 0.5. The correlation function takes the
Ornstein–Zernike form

$$g_{cc}(r) \propto \frac{1}{r} e^{-r/\xi_c} . \tag{6.64}$$

6.7   Diblock Copolymers

If n polymer species are linked together chemically one speaks of *block copolymers*
The molecules in the special case of $n = 2$ are called *diblock copolymers*. As any A
molecule is tied to a B molecule, the volume fraction of the A species is just given
by

$$\phi_A = \frac{N_A}{N_A + N_B} \tag{6.65}$$

and can only be changed by the chemist, not by the physicist. As the van-der-Waals
interactions of different polymers will still be different in the case of linking two
species together, Eq. (6.54) for the monomer-monomer interaction will hold and
one expects a segregation tendency as in the case of polymer mixtures. The overall
thermodynamics for $\chi < \chi_c$ including the critical small-angle scattering law (6.62)
will be the same. However for $\chi > \chi_c$ the molecules cannot segregate, because they
are tied together. Instead they can (and do) form *regular structures*.

These structures differ for different volume fractions. So one obtains a rich phase
diagram (see Fig. 6.4). For small concentrations $\phi_A \ll 1$ or $\phi_B \ll 1$ spheres
are formed with the minority species inside. This situation is quite comparable
to solutions of polymers with large χ parameter but with a hydrophilic end
group (lipids) in water. The spherical structures form a *bcc* lattice. For larger
fractions *cylinders* are formed, which are arranged in a hexagonal $2 - d$ lattice.

For $N_A \approx N_B$ layered structures are formed. In the concentration range $\phi \approx 0.3$ two interpenetrating diamond-type filament structures (*ordered bicontinuous double diamond, OBDD*) are formed. As the thermodynamic and interaction parameters are $A - B$ symmetric, so is the phase diagram.

From (6.60) one would estimate for $\phi_A = 0.5$, i.e., $N_A = N_B = N$ a critical χ parameter of $\chi_c = 2/N$. However empirically it was found that the transition is at $\chi \approx 5/N$.

6.8   Solid Polymers

In principle, given a polymer material with molecular weight N and a certain preferred bond-angle configuration, one can form single crystals. These crystals will have very low symmetry, due to the steric arrangements. If one has achieved to form a crystal from macromolecules one can evaluate the *molecular structure* from the Bragg line patterns. This was the method used by Watson and Krick to reveal the spiral structure of DNA and is until now the basis of biomolecular structure determination.

However, in most cases solid polymer materials are *not* crystalline. This is so, because the *crystallization kinetics* would take too much time, or, in other words, the free energy barriers between the disordered and the crystalline state are too high.

The structure of solid polymer materials is therefore highly disordered. In many cases they can be classified to be *amorphous*. One can distinguish *X-ray amorphous* from *Raman amorphous* by the method from which one has drawn the conclusion that the material is amorphous. *X-ray amorphous* material does not exhibit any Bragg peaks. Instead $S(q)$ looks quite that of a simple liquid with a peak near $q_0 \approx 2\pi/a$, where a is a mean distance between monomers. The vibration spectrum of *Raman amorphous* material has a very broad spectrum between 100 and 2,000 cm^{-1}, where no allowed vibrational Raman excitations of the crystal are present. Whereas the interpretation of the X-ray diffraction patterns is straightforward, the interpretation of the Raman spectra of glasses has only been achieved very recently [5].

In both cases one finds that *generically* solid polymer materials are in a *mixed* crystalline-amorphous state. One introduces the *crystallinity*

$$\phi_c = \frac{V_c}{V} \tag{6.66}$$

as the volume fraction of crystalline material. Quite different from crystalline materials formed by elemental atoms or small molecules crystallites are not separated by each other by grain boundaries but more or less large amorphous regions. Grain boundaries in the usual sense only exist for $\phi_c \approx 1$.

If a polymeric material is quenched from the melt it usually goes into a complete amorphous state, namely the *glassy* state. The phenomenology as well as the theoretical description of the glass transition will be subject to the lectures in the summer (Part II, section 11). Starting with the glassy *"as-quenched"* state one can try to induce relaxation processes which leads to energetically "more comfortable" positions of the monomer units. If such processes are induced by thermal treatment one calls this procedure *annealing*. They, however also occur at room temperature just as time goes on (relaxation by waiting). The result will be increasingly more and larger crystalline regions in which polymer chains are arranged parallel in a hexagonal lattice.

6.9   Gelation

We now no more consider polymers with monomer units that have bonds to $Z = 2$ nearest neighbors but *branched* macromolecules with $Z > 2$. Of course the branching may occur in reality only every N_mth monomer unit with $N_m \gg 1$. This can be easily incorporated into the considerations by rescaling the fundamental length scale.

One can take a d-dimensional lattice with coordination number Z, which is for a hypercubic lattice just $Z = 2d$ and consider the case that only a fraction of p nearest-neighbor bonds are completed. This just defines a *bond percolation model* as discussed in Sect. 5.2, if p is very small, only isolated clusters appear. A system of network-forming polymer units, in which only a small fraction has formed clusters of finite size is called a *sol*. Beyond a critical concentration p_c the network extends through the entire system and a *gel* is formed. The percolation transition in the gelation process is called *sol-gel* transition and is—as the percolation transition—a second-order phase transition, although the control parameter is not the temperature but the bond concentration. However, if one considers a bond-breaking mechanism, which is thermally activated

$$q = 1 - p \propto e^{-E_A/k_B T} \tag{6.67}$$

one has transformed the p controlled phase transition to an ordinary T controlled transition, in which the sol phase is the high-temperature phase. In other important gelation processes (e.g., rubber vulcanisation egg boiling, baking) the *bond forming* is thermally activated, which leads to gelation at high temperatures.

The first mean-field-type ideas in discussing this transition have been formulated by Flory and Stockmayer [4, 6], who considered a network without closed loops. Such a network is called a *Bethe lattice* (or *Cayley tree*, see the Fig. 6.5) with branching order (or *functionality*) Z: One starts with a point from which Z branches start. These branches lead at every vertex to $Z - 1$ further outgoing branches. The number of nodes N increases with the number n of generations as

$$\Delta N(n) = N(n) - N(n-1) = Z(Z-1)^{n-1} \tag{6.68}$$

Fig. 6.5 *Left*: Bethe lattice or Cayley tree with $Z = 3$ and $n = 4$. *Right*: Cayley tree as drawn by de Saint-Exupéry [2]

One can now consider the case in which the bonds are formed with probability p. In this case ΔN is given by

$$\Delta N(p, n) = pZ[p(Z - 1)]^{n-1} \tag{6.69}$$

If $p < p_c = 1/(Z - 1)$ the series $N(n)$ can be summed, i.e., on the average one obtains only a finite Cayley tree of size

$$\langle N \rangle = 1 + \frac{pZ}{1 - p(Z - 1)} = \frac{1 + p}{1 - p(Z - 1)} = p_c \frac{1 + p}{p_c - p}. \tag{6.70}$$

For $p \to p_c$ $\langle N \rangle$ diverges, which is then identified with the gelation threshold. As we did not make any assumption for the *angles* between successive bonds they are to be taken randomly, so that the *size* of the cluster will obey the random walk rule and is proportional to $\sqrt{\langle N \rangle}$:

$$R_Z \propto \xi \propto (p_c - p)^{-\nu} \tag{6.71}$$

with $\nu = 1/2$, which is the Flory–Stockmayer mean-field correlation length exponent.

We are now going to discuss the regime inside the gel phase. Let Q be the probability for the termination of a branch which emanates from a certain node. If the bond is absent (probability $1 - p$) $Q = 1$. If the bond is present (probability p) Q is equal to the probability for the termination of $Z - 1$ further branches. Therefore we have

$$Q = 1 - p + pQ^{Z-1}, \tag{6.72}$$

which is a closed equation for Q, albeit a nonlinear one. Obviously $Q = 1$ for $p < p_c$. For $p > p_c$ Q must be smaller than one. For $p \to p_c$ the deviation from 1 will be infinitesimally small:

$$Q = 1 - \epsilon \tag{6.73}$$

so that we can linearize (6.72):

$$1 - \epsilon = 1 - p + p[1 - (Z - 1)\epsilon] \tag{6.74}$$

Equating the coefficients of ϵ yields again $p_c = 1/(Z - 1)$.

The numerical solution of (6.72) for $Q(p)$ is depicted in the left part of Fig. 6.6. For $Z = 3$ the self consistent equation (6.72) can be evaluated analytically, as it is in this case a quadratic equation. The two solutions are

$$Q_1(p) = 1 \qquad Q_2(p) = \frac{1 - p}{p} = 1 - 2\frac{p - p_c}{p} \tag{6.75}$$

They coincide at $p = p_c = 1/2$. As Q cannot be larger than one and must be smaller than one for $p > p_c$ the physical solution is $Q_1(p)$ for $p < p_c$ and $Q_2(p)$ for $p > p_c$.

In the gel phase $p > p_c$ we can be interested in the probability P for a given node to be part of an infinite cluster. This probability together with its critical exponent β has already been introduced in the section on percolation. It is the *order parameter* of the percolation transition. In the Flory–Stockmayer Cayley-tree model this probability is, of course, zero for $p < p_c$. For $p > p_c$ there is a finite probability that all three branches terminate, which is just

$$1 - P = Q^Z \qquad \Rightarrow \qquad P = 1 - Q^Z . \tag{6.76}$$

In the case $Z = 3$ we obtain from (6.75)

$$P = 6\frac{p - p_c}{p} + O(|p - p_c|^2) , \tag{6.77}$$

i.e., $\beta = 1$. As one can see from the right part of Fig. 6.6, obviously $\beta = 1$ holds also for $Z > 3$. This result can be obtained rigorously in the following way: From (6.72) we can obtain the *inverse* of the function $Q(p) \equiv 1 - \epsilon$:

$$p(Q) = \frac{1 - Q}{1 - Q^{Z-1}} = \frac{\epsilon}{1 - (1 - \epsilon)^{Z-1}}$$

$$\approx \frac{1}{Z - 1}\frac{1}{1 - \frac{1}{2}(Z - 2)\epsilon} \approx \frac{1}{Z - 1}\left(1 + \frac{1}{2}(Z - 2)\epsilon\right) \tag{6.78}$$

From $P = 1 - (1 - \epsilon)^Z \approx Z\epsilon$ we obtain

$$P = \frac{2Z}{Z - 2}\frac{p - p_c}{p_c} , \tag{6.79}$$

which, interestingly enough, becomes independent of Z for large Z.

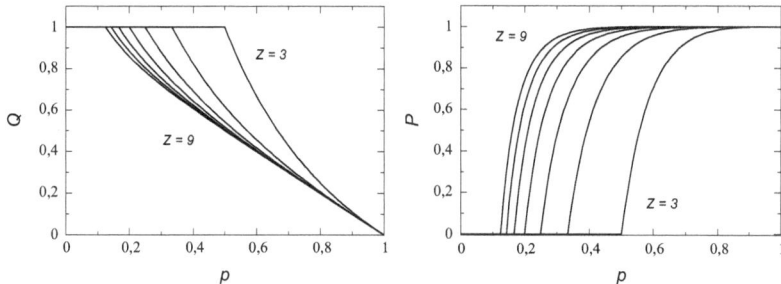

Fig. 6.6 *Left*: probability Q for termination of a branch for $Z = 3$ to $Z = 9$. *Right*: probability P for a node to be part of an infinite cluster for $Z = 3$ to $Z = 9$

The critical exponent of the mean-field theory of the Flory–Stockmayer model are, of course, quite different from those of the lattice percolation model, which includes closed loops. However, in the vulcanization transition of rubber both the chain length N_m in between the cross-linking nodes as well as the functionality Z_{eff} is so high that the effect of closed loops in the network is negligible. So the Cayley-tree model gives quite realistic results for this case. In other cases the lattice percolation theory is more adequate.

6.10   Elasticity of a Gel

We now turn to another interesting quantity, which is the elasticity of the gel phase. We here make the simplifying assumption that there is only one type of elastic response (no differentiation between longitudinal and transverse degrees of freedom) and consider the elasticity in such a way as if the mean mass m_0 of the network per node would be concentrated at the node. Then the equation of motion for the vibrations are

$$\frac{d^2}{dt^2} u_i(t) = -\sum_{j \neq i} K_{ij} [u_i(t) - u_j(t)] \equiv \sum_j D_{ij} u_j(t), \qquad (6.80)$$

with $K_{ij} = f_{ij}/m_0$ where f_{ij} are the Hookean force constants corresponding to the bond (ij) (present or absent). D_{ij} are the elements of the *dynamical matrix*, which is defined by

$$D_{ij} = \begin{cases} -\sum_m K_{im} & i = j \\ K_{ij} & i \neq j \end{cases} \qquad (6.81)$$

Table 6.1 Bond-percolation thresholds p_c compared with the Cayley-tree value $1/(Z - 1)$ and the CPA value $2/Z$

	Honeycomb	Square	Triangle	sc	bcc	fcc
Z	3	4	6	6	8	12
p_c	0.65	0.5	0.35	0.25	0.18	0.12
$1/(Z-1)$	0.5	0.33	0.2	0.2	0.14	0.09
$2/Z$	0.66	0.5	0.33	0.33	0.25	0.16

As noted already in the discussion of the vibrational properties of a fractal, Eq. (6.80) is mathematically analogous to a master equation of type (4.43), (5.15), which describes a single-particle random walk. There the double time derivative is replaced by a single one, or in frequency space the frequency parameter $-\omega^2$ is replaced by $i\omega$.

In the dc limit one obtains just the equivalent of Eq. (4.44), which means, that the determination of the elasticity of a polymer gel is mathematically equivalent to the determination of the conductance of the network with individual conductances K_{ij} [1]. At first glance one would think that the conductance of a percolating network would just follow the critical law of the order parameter. But this is not so, as noted in Sect. 5.4, because all dangling ends in the course of the gelation process contribute to the order parameter, whereas only paths which lead through the total system contribute to the conductance and the elasticity. Therefore the conductance/elasicity exponent μ (sometimes called t, but we need this letter for the *time*)

$$K \propto (p - p_c)^{\mu} \tag{6.82}$$

is larger than β and obeys the scaling relation of Gefen et al. [3]

$$\frac{\mu - \beta}{2\nu} + 1 = \frac{d_f}{d_s} = \frac{1}{\alpha} \tag{6.83}$$

In order to discuss the elasticity in the vicinity of the percolation threshold we take advantage of the mathematical analog between Eqs. (6.80) and (4.43). So we can take over the results for $\Gamma(0)$, which now has to be interpreted as the elasticity near the percolation threshold:

$$\Gamma(0) = \frac{p - p_c}{1 - p_c} \tag{6.84}$$

with the percolation threshold given in CPA $p_c = 2/Z$ and a conductivity/elasticity exponent $\mu = 1$.

It is important to note that the CPA *includes* closed loops explicitly, so that the CPA approach is complementary to the Flory–Stockmayer Cayley-tree model. We see from Table 6.1 that the critical values given in CPA compete well with the Cayley-tree ones.

References

1. de Gennes, P.G.: J. Phys. Lett. (Paris) **37**, 1 (1976)
2. de Saint-Exupéry, A.: Le Petit Prince. Gallimard, Paris (1946)
3. Gefen, Y., Aharony, A., Alexander, S.: Phys. Rev. Lett. **50**, 77 (1983)
4. Flory, P.J.: Principles of Polymer Chemistry, Cornell University Press, Ithaka and London 1953.
5. Schmid, B. Schirmacher, W.: Phys. Rev. Lett. **100**, 137402 (2008)
6. Stockmayer, W.H.: J. Chem. Phys. 11, **45** (1943)

Part II
Dynamics

Chapter 7
Time-Dependent Correlation and Response Functions

7.1 Correlation Functions

The first three sections of this chapter, which comprise the foundation of the description of liquid dynamics by means of time and space dependent correlation functions [2, 6] is formulated in the language of quantum mechanics. Afterwards we take the *classical limit*, which is accomplished by considering time scales much larger than $\tau_{\text{quantum}} = \hbar/k_B T$ or frequency scales much smaller than $\tau_{\text{quantum}}^{-1}$.

We study a many-body system with a Hamiltonian

$$\mathcal{H} = \sum_{\alpha=1}^{N} \frac{p_\alpha^2}{2m_\alpha} + \frac{1}{2} \sum_{\alpha \neq \alpha'}^{N} \phi(r_{\alpha\alpha'}), \tag{7.1}$$

where $\phi(r)$ is the pair potential, m_α are the masses of the N particles, $r_{\alpha\alpha'} = |\mathbf{r}_\alpha - \mathbf{r}_{\alpha'}|$, and $\mathbf{p}_\alpha, \mathbf{r}_\alpha$ are the quantum-mechanical momentum and position operators. If the system is in equilibrium the *canonical density operator* has the form

$$\rho = \exp\{-\beta\mathcal{H}\}/ \underbrace{\text{Tr}\{e^{-\beta\mathcal{H}}\}}_{Z}, \tag{7.2}$$

where Z is the partition function and $\beta = \hbar/k_B T$. Then the expectation value of a dynamical variable $A(t)$ is given by

$$\langle A \rangle = \text{Tr}\{\rho A\}$$
$$= \sum_i <i|\rho|i><i|A|i> \tag{7.3}$$
$$= \sum_i e^{-\beta E_i} A_{ii}$$

W. Schirmacher, *Theory of Liquids and Other Disordered Media*, Lecture Notes in Physics 887, DOI 10.1007/978-3-319-06950-0_7,
© Springer International Publishing Switzerland 2015

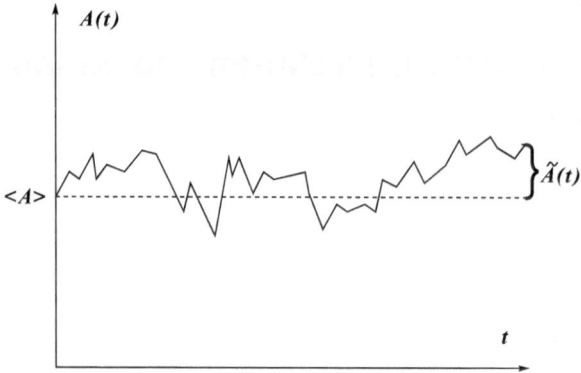

Fig. 7.1 Fluctuations of a dynamical variable $A(t)$

Here $|i >$ are the eigenstates of the Hamiltonian, E_i the eigenvalues corresponding to the characteristic equation

$$\mathcal{H}\,|i >= E_i\,|i > \tag{7.4}$$

In the present lectures the most important dynamical variable we are going to consider will be the *particle density*

$$\rho(\mathbf{r}, t) = N^{-1/2} \sum_{\alpha=1}^{N} \delta(\mathbf{r} - \mathbf{R}_\alpha(t)) \tag{7.5}$$

and its Fourier transform

$$\rho_{\mathbf{q}}(t) = \int d^3 r\, e^{i\mathbf{q}\mathbf{r}} \rho(\mathbf{r}, t) = N^{-1/2} \sum_{\alpha=1}^{N} e^{i\mathbf{q}\mathbf{R}_\alpha(t)} \tag{7.6}$$

We now want to study *fluctuations* of dynamical variables (Fig. 7.1)

$$\tilde{A}(t) = A(t) - \langle A \rangle \tag{7.7}$$

Such fluctuations can be characterized by the *van-Hove type correlation function*

$$S_{AB}(t) = \langle \tilde{A}^*(t + t_0) \tilde{B}(t_0) \rangle \tag{7.8}$$

The quantum-mechanical time dependence of $A(t)$ is given by

$$\frac{d}{dt} A(t) = i[\mathcal{H}, A(t)] \equiv i\mathcal{L}A(t), \tag{7.9}$$

where $[A, B] = AB - BA$ is the quantum-mechanical commutator and $\mathcal{L}$ is the Liouville operator. Eq. (7.9) has the solution

$$A(t + t_0) = e^{it\mathcal{L}} A(t_0) = e^{it\mathcal{H}} A(t_0) e^{-it\mathcal{H}}. \tag{7.10}$$

From the invariance of the trace with respect to cyclic permutations we deduce

$$S_{AB}(-t) = S_{B^*A^*}(t - i\beta), \tag{7.11a}$$

which implies for the density correlations

$$S_{\rho_q\rho_q}(-t) = S_{\rho_{-q}\rho_{-q}}(t - i\beta) = S_{\rho_q\rho_q}(t - i\beta) \tag{7.11b}$$

(7.11a) and (7.11b) are the *detailed-balance* relations in the time domain. We now introduce the Fourier transform

$$S_{AB}(\omega) = \int_{-\infty}^{\infty} dt\, e^{i\omega t} S_{AB}(t) \tag{7.12}$$

and, in particular

$$S(\mathbf{q}, \omega) = \int_{-\infty}^{\infty} dt\, e^{i\omega t}\, \underbrace{S_{\rho_q\rho_q}(t)}_{S(q,t)}, \tag{7.13}$$

The Fourier transform of the density-density correlation function $S(q, t)$ is related to *inelastic neutron scattering* in which between the incident beam (**k** vector $\mathbf{k}_0$) and the outgoing one (**k** vector $\mathbf{k}_1$) an amount of energy is transferred, which is given by

$$\Delta E = \hbar\omega = \frac{\hbar^2}{2m_n}\left(k_0^2 - k_1^2\right) \tag{7.14}$$

while $\hbar q = \hbar|\mathbf{k}_1 - \mathbf{k}_0|$ is, as in the static case, the momentum transfer. m_n is the neutron mass. The double-differential cross-section for scattering into a solid angle element $d\Omega$ around $\mathbf{k}_1$ per energy interval dE is given by

$$\frac{d^2\sigma}{d\Omega dE} = b^2 \frac{k_1}{k_0} S(q, \omega) \tag{7.15}$$

where b is the *scattering length*. $S(q, \omega)$ is therefore also known as *neutron scattering law*, $S(q, t)$ as *intermediate scattering function*. The corresponding cross-section for inelastic X-ray scattering with incident energy E_0, outgoing energy E_1 is given by Sette et al. [5]

$$\frac{d^2\sigma}{d\Omega dE} \propto |f(q)|^2 \frac{E_1}{E_0} S(q, \omega) \tag{7.16}$$

where $f(q)$ is the *form factor* due to the fact that X-ray scattering occurs not from the nuclei, as in the case of neutrons, but from the electronic shell of the atoms.

The initial value of $S(\mathbf{q}, t)$ is just the static structure factor $S(q)$, which has been discussed in Chap. 2.

In the frequency domain the detailed-balance relations (7.11a) and (7.11b) take the form

$$S_{AB}(-\omega) = e^{-\beta\omega} S_{B^*A^*}(\omega)$$

$$S(\mathbf{q}, -\omega) = e^{-\beta\omega} S(\mathbf{q}, \omega). \tag{7.17}$$

7.2   Linear Response and Fluctuation-Dissipation Theorem

Let us suppose we add the following perturbation to our system at $t = t_0$:

$$\delta\mathcal{H}_b(t) = -Bb(t), \tag{7.18}$$

where B is, as before, a dynamical variable. For the coupling to density fluctuations we can use a $\mathbf{q}$ dependent potential $v_\mathbf{q}$:

$$\delta\mathcal{H}_\mathbf{q} = -\rho_\mathbf{q} v_\mathbf{q} \tag{7.19}$$

By means of the time-dependent perturbation theory [4] one can show that the change in the expectation value of the variable A due to the perturbation $\delta\mathcal{H}_b$ is given by (we send $t_0 \to -\infty$)

$$\langle \delta A^*(t) \rangle = \lim_{t_0 \to -\infty} i \int_{t_0}^{t} d\tau \langle [\tilde{A}^*(t - \tau), \tilde{B}] \rangle_0 b(\tau)$$

$$\equiv \int_{-\infty}^{\infty} d\tau \chi_{AB}(t - \tau) b(\tau), \tag{7.20}$$

where $\langle \cdots \rangle_0$ denotes an average with respect to $\mathcal{H}_0$ (which is the Hamiltonian without $\delta\mathcal{H}$).

χ_{AB} is the *response function*

$$\chi_{AB}(t) = i\theta(t)\langle [\tilde{A}^*(t), \tilde{B}(0)] \rangle_0, \tag{7.21}$$

Here $\theta(t)$ is the Heaviside step function, which is defined by $\theta(t) = 1$ for $t \geq 0$ and 0 for $t < 0$. The Fourier transform of $\chi_{AB}(t)$

$$\chi_{AB}(\omega) = \int_{-\infty}^{\infty} dt\, e^{i\omega t} \chi_{AB}(t)$$

$$= \lim_{\epsilon \to 0} i \int_{0}^{\infty} dt\, e^{i\omega t} e^{-\epsilon t} \langle [\tilde{A}^*(t), \tilde{B}(0)] \rangle_0 \tag{7.22}$$

(where we inserted a convergence factor $e^{-\epsilon t}$) is the *dynamical susceptibility*. The Heaviside step function $\theta(t)$ in front of the response function (7.21) takes care of the causality, which means that there can be no answer $\langle \delta A^*(t) \rangle$ before a question $b(\tau)$ has been asked. By the convolution theorem of the Fourier transform (see Appendix A) we have

$$\delta \langle A^*(\omega) \rangle = \chi_{AB}(\omega) b(\omega), \tag{7.23}$$

which, for a magnetic perturbation $\delta \mathcal{H}_B = -MB(t)$, where M is the magnetization, reads

$$\delta \langle B^*(\omega) \rangle = \chi_{MM}(\omega) B(\omega), \tag{7.24}$$

from which the dynamic susceptibility has its name. For our density perturbation (7.19) we have

$$\delta \langle \rho^*(\omega) \rangle = \chi_{\rho_q \rho_q}(\omega) v_q(\omega), \equiv \chi(\mathbf{q}, \omega) v_q(\omega) \tag{7.25}$$

We now define a *commutator correlation function*

$$K_{AB}(t) = \langle [\tilde{A}^*(t + t_0), \tilde{B}(t_0)] \rangle_0. \tag{7.26}$$

Furthermore we define a *modified Laplace transform* (see Appendix B)

$$f(z) \equiv LT[f(t)]_z \equiv i \int_0^\infty dt\, e^{izt} f(t); \qquad \Im m\{z\} > 0 \tag{7.27}$$

so that we have

$$\chi_{AB}(\omega) = LT[K_{AB}(t)]_{z=\omega+i\epsilon} \tag{7.28}$$

From (B.12c) of the Appendix and (7.28) it follows

$$\chi''_{AB}(\omega) = \frac{1}{2} K_{AB}(\omega) = \frac{1}{2} (S_{AB}(\omega) - S_{B^*A^*}(-\omega)) \tag{7.29}$$

If we now remember the detailed-balance relation (7.17) we obtain the celebrated *fluctuation-dissipation theorem*

$$\chi''_{AB}(\omega) = \frac{1}{2} \left(1 - e^{-\beta\omega}\right) S_{AB}(\omega) \tag{7.30}$$

For the density fluctuations we have

$$\chi''(\mathbf{q}, \omega) = \frac{1}{2} \left(1 - e^{-\beta\omega}\right) S(\mathbf{q}, \omega) \tag{7.31}$$

In the classical limit $\beta\omega \ll 1$ this goes over to

$$\chi''_{class}(\mathbf{q},\omega) = \frac{\beta\omega}{2} S_{class}(\mathbf{q},\omega) \tag{7.32}$$

The inverse of relation (7.31)

$$S(\mathbf{q},\omega) = \frac{2}{1 - e^{-\beta\omega}} \chi''(\mathbf{q},\omega) \tag{7.33}$$

is the starting point of all discussions of inelastic scattering experiments. $\chi''(\mathbf{q},\omega)$ is also referred to as the excitation spectrum. The locus of the maxima of $\chi''(\mathbf{q},\omega)$ gives the dispersion relations of the quasiparticles in a solid. In a liquid it provides the dispersion of the sound modes which, however, becomes strongly broadened outside the hydrodynamic regime.

7.3 Kubo's Relaxation Function

Let us suppose that the time dependence of $b(t)$ in $\delta\mathcal{H}_b$ is of the following form

$$b(t) = \begin{cases} b_0 & t \leq t_1 \\ \\ 0 & t > t_1 \end{cases}, \tag{7.34}$$

i.e., we turn the constant field b_0 down at $t = t_1$. From time-dependent perturbation theory [4] it then follows that the quantity $\delta\langle A^*(t)\rangle$ has the following time dependence

$$\delta\langle A^*(t)\rangle = \Phi_{AB}(t - t_1)b_0 \tag{7.35}$$

with

$$\Phi_{AB}(t) = -\frac{i}{2} \int_t^\infty d\tau K_{AB}(\tau) \tag{7.36}$$

In other words, we can state

$$\frac{1}{2} K_{AB}(t) = \chi_{AB}(t) = i\frac{d}{dt}\Phi_{AB}(t) \tag{7.37}$$

Applying (B.3) of the Appendix we obtain

$$\chi_{AB}(z) = \chi_{AB} + z\Phi_{AB}(z) \tag{7.38}$$

where $\chi_{AB} = \Phi_{AB}(t = 0)$. Usually this quantity is the same as $\chi_{AB}(z = 0)$ (but see the paragraphs on non-ergodicity in Chap. 11 on the glass transition). The fluctuation-dissipation theorem can be formulated for the relaxation function in the time domain as follows

$$\Phi_{AB}(t) = \int_0^\beta d\lambda \, S_{AB}(t - i\lambda), \tag{7.39}$$

a formal relation which is only of use if one knows how to analytically continue $S_{AB}(t)$ into the complex time domain. On the other Hand, we have the following relation in the classical limit $t \gg \beta$:

$$\Phi_{AB}(t) \xrightarrow{\ t \gg \beta\ } \beta S_{AB}(t) \tag{7.40}$$

7.4 Moment Sum Rules and Continued-Fraction Expansions

We want to study the modified Laplace transform of the classical density relaxation function for a liquid[1]

$$S_{\rho_q \rho_q}(z) = \frac{1}{\beta} \Phi_{\rho_q \rho_q}(z) \equiv S(q, z) = \frac{1}{\pi} \int_{-\infty}^{\infty} d\bar{\omega} \frac{S_q''(\bar{\omega})}{\bar{\omega} - z}$$

$$= \frac{1}{2\pi} \int_{-\infty}^{\infty} d\bar{\omega} \frac{S(q, \bar{\omega})}{\bar{\omega} - z} \tag{7.41}$$

For large enough z we can make the expansion

$$S(q, z) - \frac{1}{2\pi} \frac{1}{z} \int_{-\infty}^{\infty} d\bar{\omega} \frac{S(q, \bar{\omega})}{1 - \bar{\omega}/z} = -\sum_{\nu=0}^{\infty} \frac{c_\nu(q)}{z^{\nu+1}}, \tag{7.42}$$

where $c_\nu(q)$ are the *moments* of $S(q, \omega)$

$$c_\nu(q) = \frac{1}{2\pi} \int_{-\infty}^{\infty} d\omega \, \omega^\nu S(q, \omega) = \left. (i\frac{\partial}{\partial t})^\nu S(q, t) \right|_{t=0} \tag{7.43}$$

This expansion is equivalent to a short-time expansion of $S_q(t)$ and $S(q, t)$. Because $S_q(t)$ is an even function in time, so is $S(q, t)$ in the classical limit and consequently

---

[1] As a liquid is isotropic its spatial correlations depend only on the modulus $q = |\mathbf{q}|$ of the momentum.

only the even moments are nonvanishing. For a liquid with Hamiltonian (7.1) the first moments can be calculated exactly (taking for simplicity $m_\alpha \equiv m$)

$$c_0 = S(q) = 1 + \rho_0 \int d^3r e^{i\mathbf{q}\mathbf{r}}[g(r) - 1] \tag{7.44a}$$

$$c_2 = \frac{q^2}{\beta m} \tag{7.44b}$$

$$c_4 = 3(\frac{q^2}{\beta m})^2 + \frac{\rho}{m}\frac{q^2}{\beta m} \int d^3r g(r)[1 - \cos(qz)]\frac{\partial^2 \phi(r)}{\partial z^2} \tag{7.44c}$$

On the other hand[2] $S(q, z)$ can be expanded into a continued-fraction as follows:

$$S(q, z) = \cfrac{-c_0}{z + \cfrac{-\Omega_1^2}{z + \cfrac{-\Omega_2^2}{z + \cdots}}} \tag{7.45}$$

and it is easy to show by comparing (7.42) with a $1/z$ expansion of (7.45)

$$\Omega_1^2 = c_2/c_0 = \frac{q^2}{S(q)\beta m} \tag{7.46a}$$

$$\Omega_2^2 = c_4/c_2 - c_2/c_0 \equiv \Omega_\infty^2 - \Omega_1^2 \tag{7.46b}$$

with

$$\Omega_\infty^2 = 3\frac{q^2}{\beta m} + \frac{\rho}{m} \int d^3r g(r)[1 - \cos(qz)]\frac{\partial^2 \phi(r)}{\partial z^2} \tag{7.47}$$

We can introduce residual functions, called *Memory functions* $M_\mu(z), \mu = 1, 2, \cdots$, so that we have, for example,

$$S(q, z) = \cfrac{-c_0}{z + \cfrac{-\Omega_1^2}{z + \cfrac{-\Omega_2^2}{z + M_2(z)}}} \tag{7.48}$$

The meaning of the term *Memory function* becomes clear if we recognize that

$$S(q, z) = \cfrac{-c_0}{z + \cfrac{-\Omega_1^2}{z + M_1(z)}} \tag{7.49}$$

---

[2]See next section.

is the formal solution[3] of the following differential equation for $S(q,t)$ ("generalized Langevin equation"):

$$\frac{d^2}{dt^2}S(q,t) + \int_0^t d\tau M_1(t-\tau)\frac{d}{d\tau}S(q,\tau) + \Omega_1^2 S(q,t) = 0 \qquad (7.50)$$

with the initial conditions $S(q,0) = S(q)$ and $\dot{S}(q,0) = 0$. This is a damped harmonic oscillator equation with the damping constant replaced by the memory function. The "true" damped harmonic oscillator is obtained for

$$M_1(t) = \gamma\delta(t). \qquad (7.51)$$

7.5 Projection Formalism of Mori and Zwanzig

We would like to find a way to, at least formally, give a physical meaning to the residual functions, i.e., Memory functions, introduced in the last section. This can be done by the famous projector formalism of Mori and Zwanzig [1,3,7,8].

In the present section we deal with classical correlation functions

$$\begin{aligned}
S_{AA}(t) &= \frac{1}{\beta}\Phi_{AA}(t) \\
&= \langle \tilde{A}^*(t)\tilde{A}(o)\rangle = \langle \tilde{A}^*(0)\tilde{A}(-t)\rangle \\
&= \langle \tilde{A}^* e^{-i\mathcal{L}t}\tilde{A}\rangle
\end{aligned} \qquad (7.52)$$

We now define a *scalar product*

$$(A|B) \equiv \langle \tilde{A}^*\tilde{B}\rangle \qquad (7.53)$$

The "kets" $|B)$ (and their dual "bras" $(B|$) are vectors in a generalized Hilbert space, which features operators one of which is the classical Liouville operator

$$\mathcal{L} = i\{\mathcal{H},\cdots\} \qquad (7.54)$$

where $\{ .. , .. \}$ is the Poisson bracket, the classical counterpart of the commutator [1]. As in the quantum formalism we have

$$S_{AA}(t) = \langle \tilde{A}^* e^{-i\mathcal{L}t}\tilde{A}\rangle = (A|e^{-i\mathcal{L}t}|A) \qquad (7.55)$$

---

[3]For showing this we need formula (B.7) of the Appendix.

and for the modified Laplace transform

$$S_{AA}(z) = LT[S_{AA}(t)]_z = (A|\frac{1}{\mathcal{L} - z}|A) \tag{7.56}$$

The whole idea of the projection formalism consists in classifying dynamical variables A, B and their correlation functions, determining whether or not their *static* mutual correlations are finite or not. One distinguishes between *principle* variables like density, current, pressure, which are also called *hydrodynamical* variables, and others, which are (at least statically) uncorrelated with them.

One defines a *projection operator* $\mathcal{P}_A$ as

$$\mathcal{P}_A = |A)\frac{1}{(A|A)}(A| \tag{7.57}$$

and its complement

$$\mathcal{Q}_A = 1 - \mathcal{P}_A \tag{7.58}$$

where 1 is the unit operator. In the sense of our vector calculus $\mathcal{P}_A$ projects into the direction of $|A)$ and $\mathcal{Q}_A$ projects *rectangular* to $|A)$. In what follows, we suppress the index A in $\mathcal{P}_A$ and $\mathcal{Q}_A$ to simplify the formalism.

The projectors $\mathcal{P}$ and $\mathcal{Q}$ have the properties

$$\mathcal{P}^2 = \mathcal{P} \qquad \mathcal{Q}^2 = \mathcal{Q} \qquad \mathcal{P} + \mathcal{Q} = 1 \qquad \mathcal{P}\mathcal{Q} = 0 \tag{7.59}$$

We now apply the algebraic identity

$$\frac{1}{\mathcal{U} + \mathcal{V}} = \frac{1}{\mathcal{U}}\left(1 - \frac{\mathcal{V}}{\mathcal{U} + \mathcal{V}}\right) \tag{7.60}$$

to (7.56) with $\mathcal{U} = \mathcal{L}\mathcal{Q} - z$ and $\mathcal{V} = \mathcal{L}\mathcal{P}$

$$S_{AA}(z) = (A|\frac{1}{\underbrace{\mathcal{L}\mathcal{P}}_{\mathcal{V}} + \underbrace{\mathcal{L}\mathcal{Q} - z}_{\mathcal{U}}}|A)$$

$$= (A|\left[\frac{1}{\mathcal{L}\mathcal{Q} - z} - \frac{1}{\mathcal{L}\mathcal{Q} - z}\mathcal{L}\mathcal{P}\frac{1}{\mathcal{L} - z}\right]|A) \tag{7.61}$$

The first term can be shown to give $-\frac{1}{z}(A|A)$ as follows

$$\frac{1}{\mathcal{L}\mathcal{Q} - z} = -\frac{1}{z}\frac{1}{1 - \mathcal{L}\mathcal{Q}/z} = -\frac{1}{z}[1 + (\mathcal{L}\mathcal{Q}/z) + (\mathcal{L}\mathcal{Q}/z)^2 + \cdots] \tag{7.62}$$

All the terms except the first, applied to $|A)$ give zero, which gives the desired result. Multiplying (7.61) with z we obtain

$$
zS_{AA}(z) + (A|A)
$$

$$
= (A|\frac{-z + \mathcal{L}\mathcal{Q} - \mathcal{L}\mathcal{Q}}{\mathcal{L}\mathcal{Q} - z}\mathrm{Ł}|A)\frac{1}{(A|A)}(A|\frac{1}{\mathcal{L} - z}|A)
$$

$$
= (A|\left[\mathcal{L} - \mathcal{L}\mathcal{Q}\frac{1}{\mathcal{L}\mathcal{Q} - z}\mathcal{L}\right]|A)\frac{1}{(A|A)}S_{AA}(z) \equiv [\Omega_A - M_A(z)]\,S_{AA}(z)
$$

$$
\tag{7.63}
$$

Via the transformation rule

$$
iLT[\frac{d}{dt}S_{AA}(t)] = zS_{AA}(z) + S_{AA}(t = 0) = zS_{AA}(z) + (A|A) \tag{7.64}
$$

we can re-write this equation in the time domain

$$
\frac{d}{dt}S_{AA}(t) + i\Omega_A S_{AA}(t) + \int_0^t d\tau M_A(\tau)S_{AA}(t - \tau) = 0 \tag{7.65}
$$

We call this the generalized Langevin equation of the projector formalism. The characteristic frequency is given by

$$
\Omega_A = (A|\mathcal{L}|A)/(A|A) \tag{7.66}
$$

and the memory function by

$$
M_A(z) = (A|\mathcal{L}\mathcal{Q}\frac{1}{\mathcal{Q}\mathcal{L}\mathcal{Q} - z}\mathcal{Q}\mathcal{L}|A)/(A|A) \tag{7.67}
$$

where we have repeatedly used the relation $\mathcal{Q} = \mathcal{Q}^2 = \mathcal{Q}^3, \cdots$. We now define the *generalized fluctuating force* $\mathcal{F}_A$ as

$$
\mathcal{F}_A = \mathcal{Q}\mathcal{L}|A) \tag{7.68}
$$

Obviously $M_A(t)$ is the correlation function of the fluctuating force under the action of the Liouville operator $\mathcal{Q}\mathcal{L}\mathcal{Q}$ which acts in the Hilbert space rectangular to the space spanned by $|A)$.

On the other hand we can use (7.63) to solve for $S_{AA}(z)$:

$$
S_{AA}(z) = -\frac{S_{AA}(t = 0)}{z - \Omega_A + M_A(z)} \tag{7.69}
$$

Now one can repeat the projection technique for the variable F_A:

$$M_A(z) = S_{F_A F_A}(z) = -\frac{M_A(t=0)}{z - \Omega_{F_A} + M_{F_A}(z)} \tag{7.70}$$

Inserting (7.70) into (7.69) we obtain

$$S_{AA}(z) = -\frac{S_{AA}(t=0)}{z - \Omega_A - \dfrac{M_A(t=0)}{z - \Omega_{F_A} - M_{F_A}(z)}} \tag{7.71}$$

If we now compare (7.71) with (7.45) we see a similar structure for the case that the quantities Ω_A and Ω_{F_A} vanish. This is, indeed the case for the dynamical variable $A = \rho_q$. With the help of the Mori-Zwanzig technique we are able—without making any approximations—to derive expressions for the memory functions in the continued fraction expansion (7.45).

References

1. Forster, D.: Hydrodynamic Fluctuations, Broken Symmetry and Correlation Functions. Benjamin, Reading (1975)
2. Kubo, R.: J. Phys. Soc. Jpn. **12**, 570 (1957)
3. Mori, H.: Prog. Theor. Phys. **33**, 423 (1965)
4. Schiff, L.: Quantum Mechanics. McGraw-Hill, New York (1955)
5. Sette, F., Krisch, M.H., Masciovecchio, C., Ruocco, G., Monaco, G.: Science **280**, 1550 (1998)
6. van Hove, L.: Phys. Rev. **95**, 249 (1954)
7. Zwanzig, R.: Phys. Rev. **124**, 983 (1961)
8. Zwanzig, R.: Nonequilibrium Statistical Mechanics. Oxford University Press, Oxford (2001)

Chapter 8
Collective Excitations in Simple Liquids

8.1 Linear Hydrodynamics

If we consider the dynamics of a simple liquids it is most important to first observe the conservation laws, which are valid. These conservation laws are

- particle number conservation
- momentum conservation
- energy conservation.

In the present treatment we are not interested in energy fluctuations, so we shall neglect the contributions of these quantities to the liquid dynamics.

If we denote $\rho(\mathbf{r}, t)$ the number density, $\mathbf{j}$ the corresponding current density, $\mathbf{g}(\mathbf{r}, t)$ the momentum density, τ_{ij} the components of the tensor of the momentum current density, we have from the first two conservation laws the continuity equations

$$\frac{\partial}{\partial t}\rho(\mathbf{r}, t) + \nabla \cdot \mathbf{j}(\mathbf{r}, t) = 0 \tag{8.1a}$$

$$\frac{\partial}{\partial t}g_j(\mathbf{r}, t) + \sum_i \nabla_i \tau_{ij}(\mathbf{r}, t) = 0 \tag{8.1b}$$

In addition one has the constitutive relations between $\rho, \mathbf{j}, \mathbf{g}$, and $\overset{\leftrightarrow}{\tau}$:

$$\mathbf{g}(\mathbf{r}, t) = m\mathbf{j}(\mathbf{r}, t) \tag{8.2a}$$

$$\rho_0 \tau_{im} = p(\mathbf{r}, t)\delta_{im} - \eta_S \left[\nabla_i j_m + \nabla_m j_i - \frac{2}{3} \nabla \cdot \mathbf{j}\delta_{im} \right]$$

$$- \eta_B \nabla \cdot \mathbf{j}\delta_{im} \tag{8.2b}$$

$$\nabla p(\mathbf{r}, t) = mc^2 \nabla \rho(\mathbf{r}, t) \tag{8.2c}$$

W. Schirmacher, *Theory of Liquids and Other Disordered Media*, Lecture Notes in Physics 887, DOI 10.1007/978-3-319-06950-0_8,
© Springer International Publishing Switzerland 2015

Here p is the pressure, c the sound velocity, η_S is the shear viscosity and η_B the bulk viscosity. The latter are the coefficients of internal friction for shear and dilatational (volume) distortions.

Equations (8.1) and (8.2) are the linearized Navier–Stokes equations without taking energy fluctuations into account. They are also called hydrodynamic equations as they are the basis of linear hydrodynamics. They are only valid on a length scale much larger than the atomic scale, i.e., the dynamical variables are "coarse-grained" variables, which means they are to be considered to be averaged over a meso- or macroscopic volume. "Hydrodynamic" also means that only slowly time variations of the dynamical variables are taken into account. The Brownian type irregular motion all molecules perform due to the presence of equilibrated heat motion is discarded.

We now split $\mathbf{j}$ and $\mathbf{g}$ into longitudinal and transverse contributions

$$\mathbf{g} = \mathbf{g}_\ell + \mathbf{g}_t = m(\mathbf{j}_\ell + \mathbf{j}_t) \tag{8.3a}$$

$$\nabla \times \mathbf{g}_\ell = 0 = \nabla \times \mathbf{j}_\ell \qquad \nabla \cdot \mathbf{g}_t = 0 = \nabla \cdot \mathbf{j}_t \tag{8.3b}$$

By this procedure the hydrodynamic equations (8.1) and (8.2) decouple as follows

$$\left[m\frac{\partial}{\partial t} + \frac{\eta_S}{\rho_0}\nabla^2 \right] \mathbf{j}_t(\mathbf{r}, t) = 0 \tag{8.4}$$

$$\frac{\partial}{\partial t}\rho(\mathbf{r}, t) + \nabla \cdot \mathbf{j}_\ell(\mathbf{r}, t) = 0 \tag{8.5a}$$

$$\left[m\frac{\partial}{\partial t} + \frac{1}{\rho_0}(\frac{4}{3}\eta_S + \eta_B)\nabla^2 \right] \mathbf{j}_\ell(\mathbf{r}, t) + c^2\nabla\rho(\mathbf{r}, t) = 0 \tag{8.5b}$$

Introducing now spatial and temporal Fourier transforms

$$A(\mathbf{r}, t) = \left(\frac{1}{2\pi} \right)^4 \int dt \int d^3r\, e^{-i[\omega t - \mathbf{q} \cdot \mathbf{r}]} A(\mathbf{q}, \omega) \tag{8.6}$$

we obtain from (8.4) and (8.5)

$$\left[-m\, i\omega - q^2\frac{\eta_S}{\rho_0} \right] \mathbf{j}_t(\mathbf{q}, \omega) = 0 \tag{8.7}$$

$$-i\omega\rho(\mathbf{r}, t) + i\mathbf{q} \cdot \mathbf{j}_\ell(\mathbf{q}, \omega) = 0 \tag{8.8a}$$

$$\left[-i\, m\omega + \frac{1}{\rho_0}(\frac{4}{3}\eta_S + \eta_B)q^2 \right] \mathbf{j}_\ell(\mathbf{q}, \omega) + i\mathbf{q}mc^2\rho(\mathbf{q}, \omega) = 0 \tag{8.8b}$$

Combining (8.8a) and (8.8b) we get

$$\left[\omega^2 - q^2\left(c^2 - \frac{i\omega}{m\rho_0}\eta_\ell\right)\right]\rho(\mathbf{q}, \omega) = 0 \tag{8.9}$$

where we have introduced the longitudinal viscosity

$$\eta_\ell = \frac{4}{3}\eta_S + \eta_B \tag{8.10}$$

We now introduce the correlation function of the transverse and longitudinal current

$$C_{t,l}(q, t) = \langle j_{t,l}(\mathbf{q}, t)^* j_{t,l}(\mathbf{q}, 0)\rangle \tag{8.11}$$

We have

$$C_t(\mathbf{q}, t = 0) = C_\ell(\mathbf{q}, t = 0) = c_2 = q^2\frac{k_B T}{m} \tag{8.12}$$

The equations of motion for the modified Laplace transforms of $C_t(q, t)$ and $S(q, t)$ take the form

$$\left[mz + iq^2\frac{\eta_S}{\rho_0}\right]C_t(\mathbf{q}, z) = -mC_t(\mathbf{q}, t = 0) \tag{8.13}$$

$$\left[z^2 - q^2\left(c^2 - \frac{iz}{m\rho_0}\eta_\ell\right)\right]S(q, z) = -S(q)\left(z + \frac{iz}{m\rho_0}q^2\eta_\ell\right) \tag{8.14}$$

(where we have taken $\frac{d}{dt}S(q, t)|_{t=0} = 0$) We can solve these equations to obtain

$$C_t(\mathbf{q}, z) = -\frac{k_B T/m}{z + iq^2\nu} \tag{8.15}$$

$$S(q, z) = \frac{-S(q)}{z - \dfrac{c^2 q^2}{z + iq^2\Gamma}} \tag{8.16}$$

where we have introduced the *kinematic transverse and longitudinal viscosities* $\nu = \frac{\eta_S}{m\rho_0}$, $\Gamma = \frac{\eta_\ell}{m\rho_0}$, which is at the same time the *sound attenuation constant*. We see that these expressions are the "beginning" of the continued-fraction type expressions for the correlation function. So this formalism together with the projection formalism gets the physical meaning of a *generalized hydrodynamics* as the moments and residual memory functions have a counterpart in the so-called *hydrodynamic limit $q \to 0$*.

With the help of the *dynamic susceptibility* or *response function*

$$\chi(\mathbf{q}, t) = \frac{1}{2k_B T} i \frac{d}{dt} S(\mathbf{q}, t) \tag{8.17}$$

we can re-write (8.16) as

$$\chi(\mathbf{q}, z) = \frac{1}{2k_B T} [z S(\mathbf{q}, z) + S(q)] = \frac{S(q)}{2k_B T} \left[\frac{-c^2 q^2}{z^2 + i z q^2 \Gamma - c^2 q^2} \right] \tag{8.18}$$

In all these expressions one has to use the hydrodynamic limit of the static structure factor

$$S(q) = S(q \to 0) = \rho_0 k_B T \kappa_T = \frac{k_B T}{mc^2} \tag{8.19}$$

We can also express (8.16) or (8.18) in terms of the longitudinal current correlation function, which is related to the density correlation function by the continuity equation (8.1), which gives

$$q^2 C_\ell(\mathbf{q}, t) + (\frac{d}{dt})^2 S(\mathbf{q}, t) = 0 \tag{8.20}$$

So we have (with $\frac{d}{dt} S(\mathbf{q}, t)|_{t=0} = 0$)

$$C_\ell(\mathbf{q}, z) = \frac{z}{q^2} [z S(\mathbf{q}, z) + S(q)] = \frac{2k_B T z}{q^2} \chi(\mathbf{q}, z) = z \frac{-k_B T/m}{z^2 + i z q^2 \Gamma - c^2 q^2} \tag{8.21}$$

Equations (8.15) and (8.16)–(8.21) comprise the *collective excitations* of a simple liquid in the hydrodynamic regime. The collective excitations of the transverse current fluctuations are of a *diffusive type*, which leads to a central peak in the *transverse current fluctuation spectrum*

$$C_t''(\mathbf{q}, \omega) = \text{Im}\{C_t(\mathbf{q}, z)\} = \frac{k_B T}{m} \frac{q^2 \nu}{\omega^2 + q^4 \nu^2} \tag{8.22}$$

The collective excitations of the longitudinal current fluctuations/density fluctuations are *propagating damped waves*, which lead to peaks in the current fluctuation spectrum at finite frequency $\omega = cq$ (Brillouin peak)

$$C_\ell''(\mathbf{q}, \omega) = \text{Im}\{C_\ell(\mathbf{q}, z)\} = \omega \frac{k_B T}{m} \frac{q^2 \Gamma}{(\omega^2 - c^2 q^2)^2 + \omega^2 q^4 \Gamma^2} \tag{8.23}$$

8.2 Generalized Hydrodynamics

We now come back to the continued-fraction and memory function formalism introduced in the last section. The important feature of this formalism is that before any theory is made for the memory functions, one is sure that the first important moment sum rules are automatically obeyed.

Within the continued-fraction formalism we write the transverse current and density correlation function as

$$C_t(\mathbf{q}, z) = -\frac{k_B T/m}{z + M_1^t(\mathbf{q}, z)} \tag{8.24}$$

$$S(\mathbf{q}, z) = -\frac{S(q)}{z - \dfrac{\Omega_1^2(q)}{z + M_1(\mathbf{q}, z)}} \tag{8.25}$$

These equations are the formal solutions of the following integro-differential equations (see (7.50))

$$\frac{\mathrm{d}}{\mathrm{d}t} C_t(\mathbf{q}, t) + \int_0^t \mathrm{d}\tau M_1^t(\mathbf{q}, \tau) C_t(\mathbf{q}, t - \tau) = 0 \tag{8.26}$$

$$\frac{\mathrm{d}^2}{\mathrm{d}t^2} S(\mathbf{q}, t) + \int_0^t \mathrm{d}\tau M_1(\mathbf{q}, \tau) \dot{S}(\mathbf{q}, t - \tau) + \Omega_1^2 S(\mathbf{q}, t) = 0 \tag{8.27}$$

Obviously we can state

$$\lim_{q \to 0} \mathrm{Im}\{M_1^t(\mathbf{q}, z)\} = iq^2 v \quad \lim_{q \to 0} \mathrm{Im}\{M_1(\mathbf{q}, z)\} = iq^2 \Gamma \quad \lim_{q \to 0} \Omega_1(\mathbf{q})^2 = c^2 q^2 \tag{8.28}$$

The validity of the last equality is obvious from (7.46a), whereas the two first equalities impose a limiting condition to any theory for the memory functions. As these relations are frequency independent they should be obeyed in the $\omega \to 0$ limit. So the hydrodynamic limit involves large spatial and time scales.

A *generalized hydrodynamic theory* is a theory in which theoretical expressions for the memory functions for any value of q and ω are derived or assumed which obey the relations in (8.28) are obeyed and in which the exact expressions for the frequencies $\Omega_i(q)$ in terms of the moments of the spectra are used.

Some of the existing generalized hydrodynamic approaches go one step further in the continued fraction

$$M_1(\mathbf{q}, z) = \frac{-\Omega_2^2}{z + M_2(\mathbf{q}, z)} \tag{8.29}$$

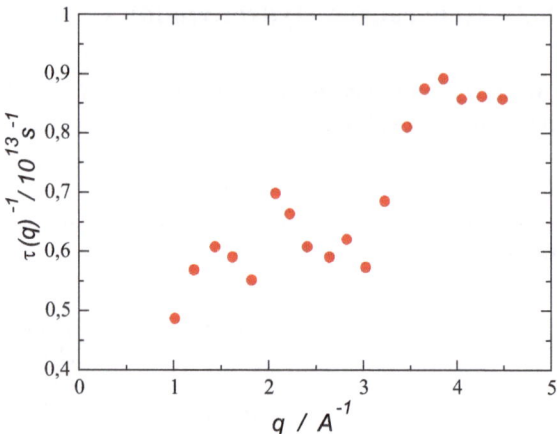

Fig. 8.1 $\tau(q)^{-1}$ as obtained by Rowe and Sköld [6] by fitting (8.25) with (8.31) to inelastic neutron $S(q, \omega)$ data for liquid Ar

so that the fourth-moment sum rule is obeyed, but use for M_2 a phenomenological Ansatz [6]

$$M_2(\mathbf{q}) = i/\tau(q) \tag{8.30}$$

which is equivalent to

$$M_1(\mathbf{q}, t) = \Omega_2^2 e^{-t/\tau(q)} \tag{8.31}$$

The function $\tau(q)$ was then determined from the measured data by fitting (Fig. 8.1).

A much more ambitious theoretical scheme for obtaining explicit expressions for the memory functions is the *mode-coupling approximation*. Such approximate schemes have been proposed both for $M_2^{\ell,t}(\mathbf{q}, t)$ and $M_1(\mathbf{q}, t)$. We shall here treat only the latter, as it has become famous in predicting and describing the *liquid-to-glass* transition (see Chap. 12), but, as we shall see in Sect. 9.4, does also a very good job in describing the collective dynamics of simple liquids away from the glass transition near the melting point.

8.3   Mode-Coupling Theory (MCT)

The idea of the mode-coupling approach [1–5] is to express the memory function $M_1(\mathbf{q}, t)$, which appears in Eq. (7.50) or (8.27) again in terms of the density correlation function $S(q, t)$. This is then a dynamic *closure relation* such as the ones for the direct correlation function in Sect. 3.7. Dealing with the memory function $M_1(\mathbf{q}, \omega)$ we have to project rectangular to the density $|\rho\rangle$ and the longitudinal current $|j_\ell\rangle(q) \equiv j_q$, so that we can write for the projector

$$\mathcal{P} = 1 - \mathcal{Q} = \frac{|\rho_q)(\rho_q|}{(\rho_q|\rho_q)} + \frac{|j_q)(j_q|}{(j_q|j_q)} \tag{8.32}$$

The memory function is given by

$$M_1(q,t) \equiv M(q,t) = \frac{m}{N}(F_j|e^{-i\mathcal{Q}\mathcal{L}\mathcal{Q}t}|F_j), \tag{8.33}$$

with the fluctuating forces

$$F_j = \mathcal{Q}\mathcal{L}j_q \tag{8.34}$$

The decoupling procedure leading to the mode-coupling mean-field equations now proceeds performing the following steps:

1. Projecting the fluctuating forces onto *pair modes of density fluctuations* $\rho_p\rho_k$.
2. Factorizing the resulting four-point density correlation functions into products of two-point functions.
3. Factorizing the static projection vertex (static three-point correlation function) into a triple product of radial pair correlation functions (Kirkwood approximation).
4. Replacing the rest of $M(q,)$, which is not proportional to products of density correlation functions by a damping term $M(q,t) = v_q\delta(t)$.

The result for $M(q,t)$ is

$$M(q,t) = \Omega_1(q)^2 m_q(t) = \Omega_1^2 \frac{1}{2V} \sum_{\substack{q+q_1+q_2=0}}^{q_1+q_2=q} V(\mathbf{q},\mathbf{q}_1,\mathbf{q}_2)S(q_1,t)S(q_2,t) \tag{8.35}$$

The vertex (in a version where the three-body static correlation functions are decoupled by the Kirkwood approximation) is given by

$$V_{\mathbf{q}\mathbf{q}_1\mathbf{q}_2} = \frac{1}{n}S(q)W_{\mathbf{q}\mathbf{q}_1\mathbf{q}_2}^2 \tag{8.36}$$

with $n = N/V$ and

$$W_{\mathbf{q}\mathbf{q}_1\mathbf{q}_2} = \frac{1}{q^2}\mathbf{q} \cdot [\mathbf{q}_1 nc(q_1) + \mathbf{q}_2 nc(q_2)] \tag{8.37}$$

where

$$c(q) = \frac{1}{n}\left(1 - \frac{1}{S(q)}\right) \tag{8.38}$$

is the direct correlation function (Sect. 2.4). The "vertex" V_{qkp} depends only on the static structure factor $S(q)$ and on the other known functions of q. Equation (8.35) together with (8.27) constitute a closed set of equations which can be solved for $S(q,t)$.

8.4 Calculation of $S(q,t)$ for Simple Liquids with MCT

Although mode-coupling theory (MCT) was devised and has been become famous for describing the liquid-to glass transition (see Chap. 11) it turned out [7, 8], that the theory in its original form [1], as described in the previous section, quite accurately describes the collective excitations in simple liquids, i.e., liquid metals. As emphasized in Sect. 3.11, the static structure factors of most simple liquids, especially those of liquid metals, can be well described in terms of the hard-sphere static structure factor in Percus–Yevick (HS-PY) approximation. However, at wavenumbers larger than the principle peak of $S(q)$ the HS-PY structure factor does not decay as rapidly towards the uncorrelated value $S(q) = 1$ as the experimental data. This corresponds to the fact that the radial pair distribution function $g(r) = 1 + n \int \frac{dq}{2\pi}^3 e^{iqr}[S(q) - 1]$ in PY approximation abruptly jumps from 0 to its maximum value at $r = \sigma$, where σ is the hard-sphere diameter. This is, of course not the case for simple liquids in which $g(r)$ rises continuously in a smooth way towards its maximum at a value of r, which is slightly larger than σ. We found that the experimentally measured $S(q)$ data of several liquid metals can be quite satisfactorily described in the total q range by the formula

$$S(q) = 1 + [S_{HS}(q) - 1]e^{-\frac{1}{2}(\lambda q \sigma)^2} \tag{8.39}$$

with $\lambda \approx 0.05$, where $S_{HS}(q)$ is the HS-PY structure factor. This corresponds to a convolution of $g_{HS}(r) - 1$ with a Gaussian of width $\lambda \sigma$ (see Fig. 8.2).

There is another, in fact, more important reason for using (8.39) instead of the HS-PY structure factor. Using the hard-sphere structure factor for calculating the memory function in (8.35) (or (11.11), resp.) leads to a vertex function $V_{qq_1q_2}$ which decays very slowly for large wavenumbers $Q = \frac{1}{2}|q_1 - q_2|$, so that there are convergence problems in the numerics.

Equations (8.27) and (8.35) have been solved by Schirmacher and Sinn [8] numerically, using the structure factor given by (8.39) with $\lambda = 0.05$, and $\eta = 0.45$ as input. As length scale we used σ as in the scaling plot of Fig. 8.2. As frequency scale we used the scale

$$\omega_0 = \frac{v_{th}}{\sigma \sqrt{S(q=0)}} = \frac{v_T}{\sigma} \tag{8.40}$$

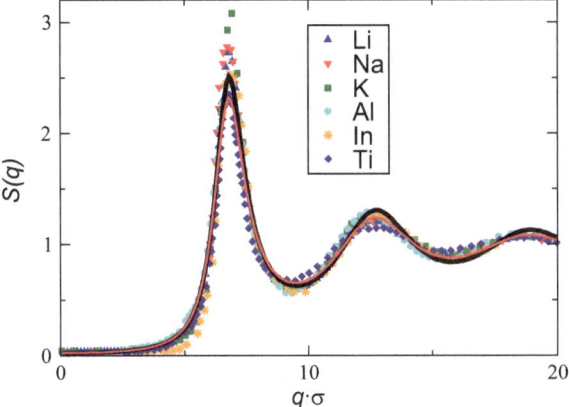

Fig. 8.2 Scaling plot for the structure factors of 6 liquid metals as in Fig. 2.5, but now including the modified HS structure factor as given by Eq. (8.39)

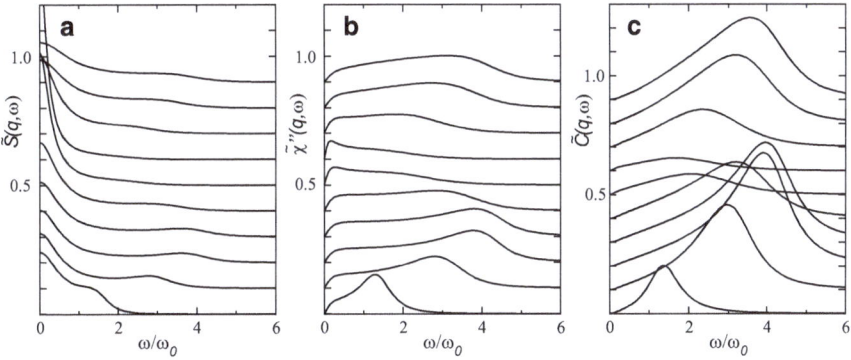

Fig. 8.3 $\tilde{S}(q,\omega) \equiv S(q,\omega)/S(q)$, $\tilde{\chi}''(q,\omega) \equiv \omega\tilde{S}(q,\omega)$ and $\tilde{C}(q,\omega) \equiv \omega^2\tilde{S}(q,\omega)$ for $\eta = 0.45$ and for (from *bottom* to *top*) $q \cdot \sigma = 1, 2, \ldots, 10$

with the isothermal sound velocity $v_T = \sqrt{1/m\rho_0\kappa_T}$. It should be noted that in these units the unrenormalized (isothermal) sound velocity is fixed to unity. This can be problematic for materials in which $S(q = 0)$ differs from the value given by HS-PY theory (see Sect. 2.12).

In Fig. 8.3 we have plotted the normalized functions $\tilde{S}(q,\omega) \equiv S(q,\omega)/S(q)$ $\tilde{\chi}''(q,\omega) \equiv \omega\tilde{S}(q,\omega) = \frac{n\pi\omega}{k_B T S(q)}S(q,\omega)$ and $\tilde{C}(q,\omega) \equiv \omega^2\tilde{S}(q,\omega)$ for integral values of $q\sigma$. It can be seen that there is both a central line and side lines that refer to collective excitations, which, for small q correspond to acoustic longitudinal waves. The dispersions (i.e., the loci of the frequency maxima of $\tilde{C}(q,\omega)$) are plotted vs. $q\sigma$ in Fig. 8.4 together with the corresponding data extracted from measured dynamic structure factors, compiled by Scopigno et al. [9]. Not only the good agreement of MCT with the measured data is remarkable, but also that the structure-factor

Fig. 8.4 Loci of the maxima of ω^2 times the measured dynamic structure factors, collected in [9], as a function of $q\sigma$, together with the corresponding curve extracted from the maxima of the $\omega^2 \widetilde{S}(q, \omega)$ curves in Fig. 8.3

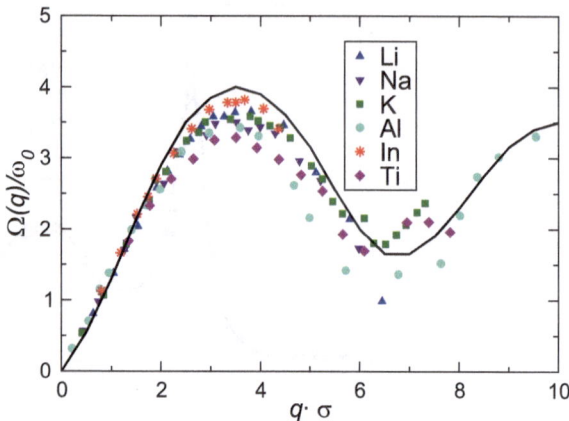

scaling obviously leads to a dispersion scaling in the collective dynamics of the liquid metals.

References

1. Bengtzelius, U., Götze, W., Sjölander, A.: J. Phys. C **17**, 5915 (1984)
2. Götze, W.: In: Hansen, J.P., Levesque, D., Zinn-Justin, J. (eds.) Liquids, Freezing and the Glass Transition. Elsevier, Amsterdam, 287 (1991)
3. Götze, W.: Condens. Matter Phys. **124**, 873 (1998)
4. Götze, W.: Complex Dynamics of Glass-Forming Liquids: A Mode-Coupling Theory. Oxford University Press, Oxford (2008)
5. Leutheusser, E.: Phys. Rev. A **29**, 2773 (1984)
6. Rowe, J.M., Sköld, K.: Neutron Inelastic Scattering. IAEA, Vienna, p. 413 (1972)
7. Said, A.H., Sinn, H., Alatas, A., urns, C.A., Price, D.L., Saboungi, M.L., Schirmacher, W.: Phys. Rev. B **74**, 892 (2006)
8. Schirmacher, W., Sinn, H.: Condens. Matter Phys. **11**, 127 (2008)
9. Scopigno, T., Ruocco, G., Sette, F.: Rev. Mod. Phys. **77**, 881 (2005)

Chapter 9
Diffusive Motion in Simple Liquids

9.1 Inelastic Neutron Scattering with a Mixture of Isotopes

It can be shown [3], that the double-differential inelastic neutron scattering cross-section from an elementary liquid with a mixture of isotopes (with different scattering lengths b_i) can be written as

$$\frac{\partial^2}{\partial\omega\partial\Omega} \propto b_{incoh}^2 S_S(\mathbf{q},\omega) + b_{coh}^2 S(\mathbf{q},\omega) \tag{9.1}$$

The incoherent and coherent squared scattering lengths are given by

$$b_{incoh}^2 = \langle b^2 \rangle - \langle b \rangle^2 \qquad b_{coh}^2 = \langle b \rangle^2 \tag{9.2}$$

where the averages are done with respect to the statistics of the isotopes. $S(\mathbf{q},\omega) = S(q,\omega)$ is the dynamical structure factor, i.e., the Fourier transform of the density-density correlation function as discussed in the previous section. Its double Fourier transform

$$G(\mathbf{r},t) = \left(\frac{1}{2\pi}\right)^4 \int \int d^3r\, dt\, e^{i[\omega t - \mathbf{q}\mathbf{r}]} S(\mathbf{q},\omega) \tag{9.3}$$

$G(\mathbf{r},t)d^3r$ can be interpreted as the probability for a particle appearing inside the volume element d^3r around $\mathbf{r}$ if there was at $t=0$ another particle (or the same) at the origin $\mathbf{r}=0$.

The quantity $S_S(\mathbf{q},\omega$ (*incoherent* scattering law) can also be represented as

$$G_S(\mathbf{r},t) = \left(\frac{1}{2\pi}\right)^4 \int \int d^3r\, dt\, e^{i[\mathbf{q}\mathbf{r} - \omega t]} S_S(\mathbf{q},\omega) \tag{9.4}$$

W. Schirmacher, *Theory of Liquids and Other Disordered Media*, Lecture Notes
in Physics 887, DOI 10.1007/978-3-319-06950-0_9,
© Springer International Publishing Switzerland 2015

$G_S(\mathbf{r}, t) d^3\mathbf{r}$ can be interpreted as the probability for a particle appearing inside the volume element $d^3\mathbf{r}$ around $\mathbf{r}$ if *it* started initially at $t = 0$ at the origin $\mathbf{r} = 0$. $S(\mathbf{q}, \omega)$ contains information about the *collective dynamics* of the liquid as discussed in the previous section, whereas $S_S(\mathbf{q}, \omega)$ describes the *individual motion* of particles in the liquid.

9.2   Individual-Particle Motion

In the *hydrodynamic*, i.e., the low-q-low-ω regime the individual motion can be described as a *random walk* as introduced in Chap. 4, i.e., $G_S(\mathbf{r}, t)$ obeys the diffusion equation

$$\frac{\partial}{\partial t} G_S(\mathbf{r}, t) = D\nabla^2 G_S(\mathbf{r}, t) \tag{9.5}$$

The solution is given in terms of the intermediate incoherent scattering function

$$S_S(\mathbf{q}, t) = \int d^3\mathbf{r}\, e^{-i\mathbf{q}\mathbf{r}} = e^{-Dq^2 t} \tag{9.6}$$

so that we have

$$S_S(\mathbf{q}, \omega) \propto \frac{Dq^2}{\omega^2 + [Dq^2]^2} \tag{9.7}$$

i.e., we have a Lorentzian spectrum with half width Dq^2. Such a scattering law is called *quasi-elastic* scattering.

We are now interested in the form of $S_S(q, \omega)$ *outside* the hydrodynamic regime.

For very large q, i.e., values much larger than $q_0 = 2\pi/a$, where a is the mean interparticle distance the motion is a *ballistic* one, i.e., a motion like in an ideal gas. This motion is characterized by a free flight with a Maxwellian distribution of velocities of variance $< v^2 >= v_{th}^2 = k_B T/m$

$$S_S(q, t) = e^{-q^2 v_{th}^2 t^2} \tag{9.8}$$

In the intermediate regime near q_0 the motion is dominated by the motion of the molecules through the "cage" of the other ones. As the particle under consideration has the same size and mass as the others, it does no more behave like a Brownian colloidal particle. This motion can occur in form of jumps (see next section), in form of double or avalanche-like multiple jumps or by "gate opening" of the cage, as has been studied in detail by means of molecular-dynamics simulations.

Fig. 9.1 The Chudley–Elliott
function $f(q)$

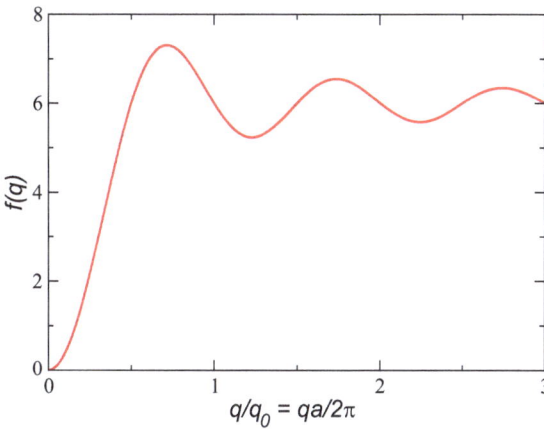

9.3   Jump Diffusion

It has been suggested by Chudley and Elliot [1] that this motion may take place in
steps of size a like a random walk on a lattice. As shown in Sect. 5.3 this random
walk can be characterized by

$$S_S(\mathbf{q}, z) = -\frac{1}{z + if(\mathbf{q})} \tag{9.9}$$

with

$$f(\mathbf{q}) = 2W[3 - \cos(q_x a) - \cos(q_y a) - \cos(q_z a)] \tag{9.10}$$

where W is the jump rate. In the hydrodynamic regime we have

$$f(\mathbf{q}) = Wa^2 q^2 = Dq^2 , \tag{9.11}$$

so that (9.9) reduces to the ordinary diffusion propagator. Performing an angular
average we obtain the Chudley–Elliott half-width function (Fig. 9.1).

$$f(q) = \langle f(\mathbf{q}) \rangle_{\text{angles}} = 6W[1 - \sin(qa)/qa] \tag{9.12}$$

and we have an incoherent scattering law of the form

$$S(q, \omega) \propto \frac{f(q)}{\omega^2 + f(q)^2} \tag{9.13}$$

where, again, $f(q) \to Dq^2$ for $q \to 0$.

This is, of course, a very schematic way of describing the microscopic particle motions. The only important conclusion by this model calculation is that outside the hydrodynamic regime the width of the quasielastic line will certainly deviate from the q^2 behavior and will eventually level off in some way. Such a behavior is, in fact, observed in a number of liquids, giving evidence for a jump-wise diffusion process. From the first maximum of the width of the quasielastic line (plotted against q) the spatial extension a of the jumps can be extracted [2].

9.4 The Diffusivity of Interacting Colloidal Particles

In colloidal suspensions particles of diameter of several 100 nanometers are suspended in a solvent. Due to the interaction with the solvent the particles perform a diffusive motion. Therefore the dynamic structure factor of the ensemble of the particles (which can be measured with inelastic light scattering) is given by

$$S(q,\omega) = \frac{1}{\pi} S(q) \frac{D_c q^2}{\omega^2 + [D_c q^2]^2} \tag{9.14}$$

which is equivalent to

$$S(q,z) = \frac{-S(q)}{z + i D_c q^2} \tag{9.15}$$

The dynamic susceptibility[1] is given by

$$\tilde{\chi}(q,z) = -\rho_c \chi(q,z) = -\rho[S(q) + zS(q,z)] = -\frac{\rho_c}{k_B T} \frac{i D_c q^2}{z + i D_c q^2} \tag{9.16}$$

where $S(q)$ is the static structure factor of the colloidal particles and D_c the collective diffusivity.

$$\tilde{\chi}(q,z) = -\frac{\rho_c}{k_B T} S(q) \frac{i D_c q^2}{z + i D_c q^2} \tag{9.17}$$

The self motion of the colloid, on the other hand, can be described as in a simple liquid by

$$S_S(q,z) = \frac{-1}{z + i D_S q^2} \tag{9.18}$$

---

[1]This is the definition in the non-MCT literature.

and

$$\tilde{\chi}_S(q,z) = -\frac{\rho_c}{k_B T} \frac{i D_S q^2}{z + i D_S q^2} \tag{9.19}$$

The structure factor can—as in a simple liquid—be written in terms of a direct correlation function $c(q)$

$$S(q) = \frac{1}{1 - \rho_c c(q)} = \frac{1}{1 + \frac{\rho_c}{k_B T} U_{\text{eff}}(q)} \tag{9.20}$$

where ρ_c is the number density of the particles.

We are now going to generalize the RPA for the dynamic susceptibility $\tilde{\chi}(q,z)$,

$$\tilde{\chi}(q,z) = \frac{\tilde{\chi}_S(q,z)}{1 - U_{\text{eff}}(q) \tilde{\chi}_S(q,z)} \tag{9.21}$$

which can be put into the form, using the χ instead of the $\tilde{\chi}$

$$\frac{k_B T}{\chi(q,z)} - \frac{k_B T}{\chi_S(q,z)} = \frac{\rho_c}{k_B T} U_{\text{eff}}(q) = -\rho_c c(q) = \frac{1}{S(q)} - 1 \tag{9.22}$$

We now utilize the continued-fraction representations of $S(q,z)$ and $S_S(q,z)$

$$S(q,z) = \frac{-S(q)}{z - \dfrac{q^2 k_B T / m S(q)}{z + M(q,z)}}; \qquad S_S(q,z) = \frac{-1}{z - \dfrac{q^2 k_B T / m}{z + M_S(q,z)}} \tag{9.23}$$

from which follows

$$\frac{k_B T}{\chi(q,z)} = \frac{1}{S(q)} - \frac{z[z + M(q,z)]}{q^2 k_B T / m}; \qquad \frac{k_B T}{\chi_S(q,z)} = 1 - \frac{z[z + M_S(q,z)]}{q^2 k_B T / m} \tag{9.24}$$

Taking the difference we see that the RPA is equivalent to setting the two memory functions $M(q,z)$ and $M_S(q,z)$ equal to each other. If we compare (9.15), (9.18) with (9.23) we can define generalized wavevector and frequency dependent diffusivities

$$D_c(q,z) = i \frac{k_B T / m S(q)}{z + M(q,z)}; \qquad D_S(q,z) = i \frac{k_B T / m}{z + M_S(q,z)} \tag{9.25}$$

Of course in the $q \to 0$ and $z \to 0$ limit these function must become equal to the hydrodynamic diffusivities D_c and D_S. If we now set the memory functions M and M_S equal to each other we obtain the RPA expression for the collective hydrodynamic diffusivity

$$D_c = \frac{1}{S(q \to 0)} D_S = \frac{1}{\rho_c k_B T \kappa_T} D_S = f_c D_S \tag{9.26}$$

where κ is the isothermal compressibility of the colloid particles,

$$\kappa_T = -\frac{1}{V_c}\left(\frac{\partial V_C}{\partial \Pi}\right)_T \tag{9.27}$$

where V_C is the volume occupied by the colloid particles and Π is the osmotic pressure. The correlation factor $f_c = 1/S(0)$ can be rather large, because the colloid structure factor can be well approximated by a hard-sphere structure factor, which at $q = 0$ has values of the order of 0.02 for packing fractions near $\eta = 0.45$ (see Sect. 2.12).

References

1. Chudley, C.T., Elliot, R.J.: Proc. Phys. Soc. **77**, 353 (1961)
2. Faupel, F., Frank, W., Macht, M.-P., Mehrer, H., Naundorf, V., Rätzke, K., Schober, H.R., Sharma, S.K., Teichler, H.: Rev. Mod. Phys. **75**, 237 (2003)
3. Lovesey, S.W.: Theory of Neutron Scattering from Condensed Matter. Clarendon Press, Oxford (1987)

Chapter 10
Polymer Dynamics

10.1 Dynamics of a Single Polymer: Rouse Model

We now would like to derive a simple model for the dynamics of an ideal polymer chain, i.e., the mean conformational changes as a function of time. The chain conformation is given by N monomer vectors $\mathbf{r}_i(t)$, $(i = 1 \cdots N)$ which point along the chain and have the length a of a bead.

The elastic free energy of the chain is given by (see Chap. 6)

$$F_{\text{el}} = \frac{1}{2} K |\mathbf{r}_N - \mathbf{r}_1|^2 \qquad (10.1)$$

with the global force constant

$$K = \frac{3 k_B T}{R_0^2} \qquad (10.2)$$

where $R_0 = \sqrt{N} a$ is the radius of gyration. Every monomer shares $1/N$ of this force, so that one can write down with $k = K/N = 3 k_B T / a^2$

$$m \ddot{\mathbf{r}}_n(t) = -k \sum_{\ell \neq n} (\mathbf{r}_n(t) - \mathbf{r}_\ell(t)) - \alpha_H \dot{\mathbf{r}}_n(t) \qquad (10.3)$$

Here m is the mass of a monomer and we have introduced the hydrodynamic friction with the coefficient

$$\alpha_H = 6 \pi \eta_S R_H \qquad (10.4)$$

W. Schirmacher, *Theory of Liquids and Other Disordered Media*, Lecture Notes in Physics 887, DOI 10.1007/978-3-319-06950-0_10,

where η_S is the viscosity and R_H the so-called hydrodynamic radius, which controls the effective Einstein relation between the diffusivity D_H of a part of the polymer and η_S

$$D_H = \frac{k_B T}{6\pi \eta_S R_H} \tag{10.5}$$

We now assume that the elastic and friction forces are much stronger than the inertial forces $f_n = m\ddot{\mathbf{r}}_n$. If we further assume that the elastic forces act only among nearest neighbors we obtain

$$\dot{\mathbf{r}}_n(t) = W\ (\mathbf{r}_{n+1} + \mathbf{r}_{n-1} - 2\mathbf{r}_n) \tag{10.6}$$

with the "hopping" rate

$$W = \frac{3D_H}{a^2} \tag{10.7}$$

If we now introduce a coordinate ξ along the chain and go into the continuum limit $a \to 0$ we obtain

$$\dot{\mathbf{r}}(\xi,t) = \underbrace{3D_H}_{\tilde{D}}\ \frac{\partial^2}{\partial\xi^2}\mathbf{r}(\xi,t) \tag{10.8}$$

We see that a given "excitation", described by $\mathbf{r}(\xi,t)$ performs a diffusive motion along the chain. One can think of a kink produced at time $t = 0$ in the middle of the chain at a chain position $\chi = \chi_0$. If the chain is very long, the probability $P(\xi,t)$ to find an excitation caused by the perturbation is for times, before the excitation becomes aware that the chain is finite:

$$P(\xi,t) = \frac{1}{4\pi\tilde{D}T}e^{-(\xi-\xi_0)^2/4\tilde{D}t} \tag{10.9}$$

For the chain ends we have the boundary conditions

$$\dot{P}(\xi = 0, t = 0) = \dot{P}(\xi = L, t = 0) = 0 \tag{10.10}$$

and $P(\xi,t)$ obeys the same equation of motion as $\mathbf{r}(\xi,t)$

$$\dot{P}(\xi,t) = \underbrace{\tilde{D}}_{\tilde{W}a^2}\ \frac{\partial^2}{\partial\xi^2}P(\xi,t) \tag{10.11}$$

To solve this equation with the boundary conditions (10.10) we make the separation ansatz

$$P(\xi,t) = u(t)w(\xi) \tag{10.12}$$

If we multiply the "diffusion wave equation" (10.11) from the left with $1/u(t)w(\xi)$ we obtain

$$\frac{1}{Wu(t)}\frac{\partial}{\partial t}u(t) = \frac{a^2}{v(\xi}\frac{\partial^2}{\partial \xi^2}w(\xi) = -\lambda^2 \tag{10.13}$$

$-\lambda^2$ must be a constant because the term on the left of the first equality sign depends only on the time, whereas the term on the right of the first equality sign depends only on ξ. For the function $u(t)$ we obtain the differential equation

$$\frac{d}{dt}u(t) = -\lambda^2 Wu(t) \tag{10.14}$$

which has the general solution

$$u(t) = u(0)e^{-\lambda^2 Wt} \tag{10.15}$$

For the function $w(\xi)$ we obtain the second-order differential equation

$$\left(\frac{d^2}{d\xi^2} + \frac{\lambda^2}{a^2}\right)w(\xi) = 0 \tag{10.16}$$

which has the general solution

$$w(\xi) = A\cos(k\xi) + B\sin(k\xi) \tag{10.17}$$

with

$$k = \frac{\lambda}{a} \tag{10.18}$$

The first boundary condition in (10.10) excludes the sine term. From the second boundary condition we find that only discrete values of k and λ are possible

$$k_v = \frac{\lambda_v}{a} = \frac{\pi v}{L} \quad \Leftrightarrow \quad \lambda_v = \frac{\pi v}{N} \qquad v = 1, 2 \cdots N \tag{10.19}$$

For the general solution including the boundary conditions we obtain

$$P(\xi, t) = A_v u_v(0)e^{-\lambda_v^2 Wt}\cos(k_v\xi) \tag{10.20}$$

which has the modified Laplace transform (see Appendix B)

$$P_v(\xi, z) = -\frac{A_v u_v(0)\cos(k_v\xi)}{z + i\lambda_v^2 W} \tag{10.21}$$

The *relaxation spectrum* is given by

$$\frac{1}{\pi} \text{Im} \{\phi(z)\} = \frac{1}{\pi} \text{Im} \left\{ \sum_\nu \frac{-1}{z + i\lambda_\nu^2 W} \right\} = \frac{1}{\pi} \sum_\nu \frac{1}{1 + \omega^2 \tau_\nu^2} \tag{10.22}$$

with the *mode relaxation times*

$$\frac{1}{\tau_\nu} = \lambda_\nu^2 W = \left(\frac{\nu\pi}{N}\right)^2 W \tag{10.23}$$

The dynamic susceptibility is given by

$$\chi(z) = 1 + z\phi(z) = \sum_\nu \frac{1}{1 - i\omega\tau_\nu} \tag{10.24}$$

from which follows

$$\chi''(\omega) = \sum_\nu \frac{\omega\tau}{1 + \omega^2 \tau \nu^2} \tag{10.25}$$

The Rouse model suffers from being too "mean-field like", i.e., it does not contain the effect of interactions between the monomers, except the elastic spring constants.

10.2   Rouse Dynamics with a Distribution of Interaction Constants

The unrealistic feature of the equations of motion of the rouse model is not the mathematical structure but the assumption *that every bead is subject to the same transition rate W*. In reality this is not so because of the chain-chain interactions. To include these interactions one must in principle deal with *non-linear equations* which are usually very difficult to solve. Instead we shall introduce a *statistical distribution* of transition rates and write

$$\dot{P}_i(t) = W_{i,i+1}(P_{i+1} - P_i) - W_{i,i-1}(P_{i-1} - P_i) \tag{10.26}$$

or, for the Laplace transform with Laplace parameter $s = -i\omega$

$$sP_i(t) - \delta_{ii_0} = W_{i,i+1}(P_{i+1} - P_i) - W_{i,i-1}(P_{i-1} - P_i) \tag{10.27}$$

where we imposed the initial condition

$$P_i(t = 0) = \delta_{ii_0} \tag{10.28}$$

If we now define the quantities

$$g_i = W_{i,i+1}\left(\frac{P_i - P_{i+1}}{P_i}\right) \tag{10.29}$$

we obtain for $i, i+1 \neq i_0$ the recursion relation [1]

$$g_i = \frac{1}{\dfrac{1}{W_{i,i+1}} + \dfrac{1}{g_{i+1} + s}} \tag{10.30}$$

This equation can be easily iterated in a computer simulation for a given distribution of rates $W_{i,i+1}$.

Instead of such a simulation we shall, instead, again use a mean field approach, which is the CPA introduced Sect. 4.4. There we considered as effective medium a simple cubic lattice in $d = Z/2$ dimension, where Z is the coordination number. Taking $Z = 2$ we obtain from (4.53)

$$\langle\frac{W - \Gamma(s)}{1 + (W - \Gamma(s))\frac{1}{\Gamma(s)}(1 - sG_{ii}(s))}\rangle = 0 \tag{10.31}$$

As we now do *not* deal with a stretched-out chain but a wrinkled polymer chain, we cannot take (4.54) for the local Green's function G_{ii}. Instead we just take the Green's function of an effective medium, where only hops *away* from a given site i is possible:

$$G_{ii}(s) = \frac{1}{s + \Gamma(s)} \tag{10.32}$$

From these equations we obtain the simple equation

$$\langle G_{ii}(s)\rangle = \frac{1}{s + \Gamma(s)} = \langle\frac{1}{s + W}\rangle \tag{10.33}$$

This equation can be interpreted in the following way: Let us consider a chain of conductances $W_{i,i+1}$ in series the leads of which are grounded with unit capacitance. The set of Kirchhoff's equations of such a circuit network is just the set (10.27). The complex impedance of the network is then given by $s + \Gamma$ which is the serial sum given by (10.33). In particular the dc limit of (10.33) is given by the serial formula

$$\frac{1}{\Gamma(0)} = \langle\frac{1}{W}\rangle \tag{10.34}$$

We now introduce a model for a distribution of transfer rates W, which is very common in the literature, which deals with structural relaxation of soft matter.

We assume that each unit has to overcome a certain *free energy barrier* E to go to its neighboring conformational state:

$$W_{ij} = v_0 e^{-E/k_B T} \tag{10.35}$$

and assume that the barriers are distributed according to a distribution density $g(E)$. To be specific we assume

$$g(E) = \frac{1}{E^*} \theta(E^* - E) \tag{10.36}$$

i.e., a constant distribution with a cutoff E^* in order to have a normalized distribution.

The average in (10.33) can be easily performed to obtain

$$\Gamma(s) = s \frac{k_B T}{E^*} / \ln \left[\frac{s + v_0 e^{-E^*/k_B T}}{s + v_0} \right] - s \tag{10.37}$$

For $s = 0$ we obtain (using l'Hopital's rule)

$$\Gamma(0) = \frac{k_B T}{E^*} \frac{v_0}{e^{E^*/k_B T} - 1} \approx \frac{k_B T}{E^*} v_0 e^{-E^*/k_B T} \tag{10.38}$$

We now introduce *scaled variables*

$$\tilde{s} = s/\Gamma(0) \qquad \tilde{\Gamma}(s) = \Gamma(s)/\Gamma(0) \tag{10.39}$$

where we take the second expression for $\Gamma(0)$ in (10.38).

$$\tilde{\Gamma}(s) = \tilde{s} \left(\frac{k_B T/E^*}{\ln(\tilde{s} + 1) - \ln(\tilde{s} e^{-E^*/k_B T} + 1)} - 1 \right) \tag{10.40}$$

For all frequencies smaller than $v_0 \approx 1\,\mathrm{THz}$ the second term in the denominator of (10.40) is negligible and we obtain

$$\tilde{\Gamma}(s) = \tilde{s} \left(\frac{k_B T/E^*}{\ln(\tilde{s} + 1)} - 1 \right) \tag{10.41}$$

We see that the dynamics obeys a scaling law, i.e., the time dependence of the generalized scaled relaxation rate $\tilde{\Gamma}(\tilde{s})$ is completely determined by the universal function (10.41), which is depicted in Fig. 10.1. Such a scaling law is called *time-temperature scaling law*.

This law is also obeyed by the *dynamic susceptibility*

$$\tilde{\chi}(\tilde{s}) = \frac{(\tilde{\Gamma}(\tilde{s})}{\tilde{\Gamma}(\tilde{s}) + \tilde{s}} \tag{10.42}$$

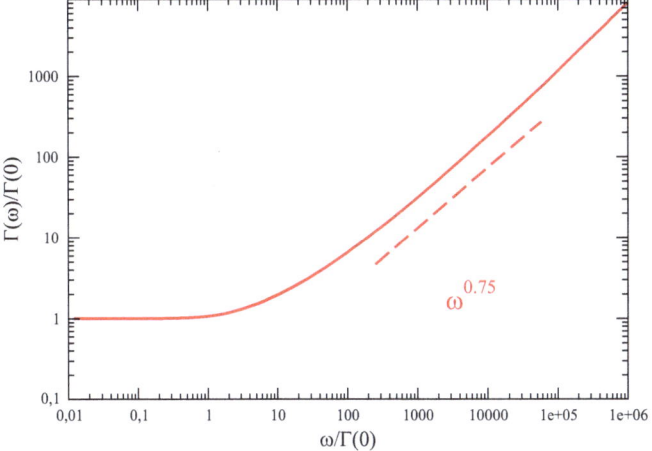

Fig. 10.1 Real part of the "self energy" $\Gamma(\tilde{z})$ for the random barrier model (10.37)

From Fig. 10.1 we see that for $\tilde{s} \gg 1$ the universal function can be approximately described by

$$\tilde{\Gamma}(\tilde{s}) \approx \tilde{A}\tilde{s}^n \tag{10.43}$$

with $n \approx 0.75$. For the dynamic susceptibility we obtain

$$\tilde{\chi}(\tilde{s}) \approx \frac{1}{1 + \tilde{A}^{-1}\tilde{s}^{1-n}} \tag{10.44}$$

Such a frequency dependence is frequently found empirically for the dynamics of soft matter and is called *Cole-Cole* behavior. It is revealing to compare the original Rouse expression for $\chi(\omega)$ with our model expression

$$\chi(z) = \sum_\nu \frac{1/\tau_\nu}{-i\omega + 1/\tau_\nu}1 \qquad \chi(z) = \int g(E)dE \frac{W(E)}{-i\omega + W(E)} \tag{10.45}$$

The mean-field expression for the dynamic susceptibility is obviously just the *average* of the single-bead susceptibilities. The integral over the continuous distribution of the relaxation rates removes the artificial discreteness of the Rouse dynamics. Of course, for large ν the Rouse spectrum is almost continuous, but not for small ν. Moreover in the Rouse model the aspect of *activated barrier motion* induced by the inter-chain interactions is not taken into account.

The present model treats the inter-chain interactions as a source of static disorder. If one wants to take the dynamics into account in a self-consistent way one must utilize a suitable version of mode-coupling theory.

10.3   Incoherent Relaxation Dynamics

We now treat a phenomenological model for the relaxation dynamics of soft matter which is very popular in the community dealing with dielectric and optical response techniques [2].

One assumes that the relaxation function of the material is an *incoherent sum* of individual relaxation functions with a relaxation time characterized by an activation free energy E (as in the generalized Rouse model)

$$\phi(z) = \int dE g(E) \frac{-1}{z + i/\tau(E)} \tag{10.46}$$

with

$$\frac{1}{\tau(E)} = \nu_0 e^{-E/k_B T} \tag{10.47}$$

Different from the generalized Rouse model we now leave the distribution of activation energies $g(E)$ unspecified. The quantity of interest in dielectric or optical absorption measurements is the imaginary part of the dynamical susceptibility (loss function, out-of-phase susceptibility)

$$\chi''(\omega, T) = \omega \mathrm{Im}\{\phi(z = \omega + i\epsilon)\} = \int dE g(E) \frac{\omega \tau(E)}{1 + [\omega \tau(E)]^2} \tag{10.48}$$

we now define a characteristic energy E_ω as

$$E_\omega = -k_B T \ln(\omega/\nu_0) \tag{10.49}$$

which, of course, is positive, because $\omega \ll \nu_0$. We obtain for the loss function

$$\chi''(\omega, T) = \int dE g(E) \frac{e^{[E - E_\omega]}}{1 + e^{2[E - E_\omega]}} \tag{10.50}$$

Because the exponential varies very rapidly with energy we can write

$$\frac{e^{[E - E_\omega]}}{1 + e^{2[E - E_\omega]}} \approx \frac{1}{\pi k_B T} \delta(E - E_\omega) \tag{10.51}$$

from which follows

$$\chi''(\omega, T) \approx \frac{1}{\pi k_B T} g(E_\omega) \tag{10.52}$$

This means that if one can justify the assumption of individual relaxation processes (which is not always the case) then one can "measure" the distribution $g(E)$ from the temperature dependence of the out-of-phase susceptibility.

10.4 Hydrodynamic Interaction

One type of the interactions, which are present in a polymer solution (as well as in any solutions of macromolecules) is the interaction which is mediated by the solvent. As we describe the motions of the solute in terms of the local velocity of the segments we need to know the effect of a force on the velocity field of the solvent. Treating the solvent as incompressible and assuming that all the velocity excitations are transverse we have the Navier–Stokes equation

$$\rho_m \dot{\mathbf{v}} - \eta_s \nabla \nabla \cdot \mathbf{v} = \mathbf{f} \qquad\qquad \nabla \cdot \mathbf{v} = 0 \qquad\qquad (10.53)$$

Here ρ_m is the solvent mass density and $\mathbf{f}$ is an external force. The solution of (10.53) in wavevector and frequency space is given by

$$\mathbf{v}(\mathbf{q}) = \frac{1}{-i\omega\rho_m + \eta_s q^2}\left[1 - \mathbf{q}\frac{1}{q^2}\mathbf{q}\cdot\right]\mathbf{f} \equiv T(p, \mathbf{q})\mathbf{f} \qquad (10.54)$$

The projector inside the square bracket projects rectangular to $\mathbf{q}$ to take care of the condition $\mathbf{q} \cdot \mathbf{v} = 0$.

If we compare (10.54) with (8.15) we see that the matrix elements of the tensor T are proportional to the Laplace transforms of the diagonal and off-diagonal current-current correlation functions of the solvent:

$$T_{\alpha\beta}(\omega, \mathbf{q}) = k_B T \rho_0 \int_0^\infty dt\, e^{[i\omega - \epsilon]t} \langle j_\alpha(\mathbf{q}, t + t_0) j_\beta(\mathbf{q}, t)\rangle \qquad (10.55)$$

where ρ_0 is the number density of the solvent. The time integral over $T(t, \mathbf{q})$, i.e., $T(\omega = 0, \mathbf{q}) \equiv T(\mathbf{q})$ is the so-called *Oseen tensor*, which has the real-space representation

$$T(\mathbf{r}) = \frac{1}{8\pi\eta_s r}\left[1 + \mathbf{r}\frac{1}{r^2}\mathbf{r}\cdot\right] \qquad (10.56)$$

so that the velocity response at $\mathbf{r}$ due to a force at $\mathbf{r}'$ is given by

$$\mathbf{v}(\mathbf{r}) = \int d^3\mathbf{r}'\, T(\mathbf{r}' - \mathbf{r})\mathbf{f}(\mathbf{r}') \qquad (10.57)$$

We see that the hydrodynamic interaction $T(\mathbf{r} - \mathbf{r}')$ is a *long-range* interaction like the Coulomb interaction.

10.5   Zimm Model

In order to incorporate the important hydrodynamic forces into the Rouse model Zimm[1] inserted the Oseen tensor into the dynamics in the following way:

$$\dot{\mathbf{r}}_i = -\sum_j \mathcal{H}_{ij}(\mathbf{r}_i - \mathbf{r}_j) \tag{10.58}$$

where $\mathcal{H}_{ij}(\mathbf{r})$ is the *mobility matrix*

$$\mathcal{H}_{ij}^{\alpha\beta}(\mathbf{r}) = W \delta_{ij}\delta\alpha\beta + (1 - \delta_{ij})\mathcal{T}_{\alpha\beta}(\mathbf{r}) \tag{10.59}$$

The Zimm equation (10.58) is a *nonlinear* equation for the $\mathbf{r}_i(t)$ and can only be solved numerically. If in a so-called *pre-averaging approximation* the Oseen tensor is replaced by its thermal average one retains a linear problem. The pre-averaged Zimm dynamics has, as the Rouse dynamics, a *discrete* relaxation spectrum.

10.6   Diffusivity of a Single Polymer Chain in Solution

We now want to apply the Kubo relation (C.11) to the diffusivity of a single chain in solution. We represent the polymer velocity as

$$\mathbf{V}(t) = \frac{1}{N}\sum_n \mathbf{v}_n \tag{10.60}$$

and identify the local velocity $\mathbf{v}_n(t)$ with the local solvent velocity times the monomer concentration at $\mathbf{r}_n$:

$$\mathbf{v}(\mathbf{r}_n, t) = c(\mathbf{r}_n, t)\mathbf{v}(\mathbf{r}_n, t) \tag{10.61}$$

so that we have

$$\mathbf{V}(t) = \frac{1}{N}\int d\mathbf{r} c(\mathbf{r}, t)\mathbf{v}(\mathbf{r}, t) \tag{10.62}$$

We now insert this into the Kubo relation and obtain

$$D = \frac{1}{N^2}\int d^3\mathbf{r}_1 \int d^3\mathbf{r}_2 \int_0^\infty dt \langle c(\mathbf{r}_1, t+t_0)c(\mathbf{r}_2, t)\mathbf{v}(\mathbf{r}_1, t+t_0)\mathbf{v}(\mathbf{r}_2, t)\rangle \tag{10.63}$$

---

[1]B.H. Zimm, J. Chem. Phys. **24**, 269 (1956).

Now two simplifications are done: First the averages over the concentrations and those over the velocities are decoupled, so that we obtain the product of the concentration correlation function and the solvent velocity correlation function. Secondly the time dependence of the concentration correlation function is neglected. For the concentration correlation function one takes the polymer pair correlation function $\tilde{g}(\mathbf{r})$ (6.22) of Sect. 7.1. We finally arrive at

$$D = \frac{k_B T}{3N} \int d^3\mathbf{r}\, \tilde{g}(\mathbf{r}) \sum_\alpha T_{\alpha\alpha}(\mathbf{r}) = \frac{1}{N} \int d^3\mathbf{r}\, \tilde{g}(\mathbf{r}) D_H(\mathbf{r}) \qquad (10.64)$$

with the hydrodynamic diffusion coefficient

$$D_H(\mathbf{r}) = \frac{k_B T}{6\pi \eta_s r} \qquad (10.65)$$

We now remind ourselves the scaling law for $\tilde{g}(\mathbf{r})$

$$g(\mathbf{r}) = \frac{N}{R^3} f\left(\frac{r}{R}\right) \qquad (10.66)$$

Here $R = R_0 \propto N^{1/2}$ in a Θ solution and $R = R_F \propto N^{3/5}$ for a dilute solution. From this we obtain for the diffusivity

$$D = \frac{k_B T}{6\pi \eta_s R} \int_0^\infty dx f(x) x = C_0 \frac{k_B T}{6\pi \eta_s R} \qquad (10.67)$$

So the polymer diffuses as if it were a colloidal sphere with effective radius $R_{rmeff} = R/C_0$.

References

1. Alexander, S., Bernasconi, J., Schneider, W.R.: Rev. Mod. Phys. **53**, 175 (1981)
2. Wong, J., Angell, C.A.: Glass: Structure by Spectroscopy. Marcel Dekker, New York (1976) [ISBN 0824764684]

Chapter 11
Glass Transition and Glass Dynamics

11.1   Non-ergodicity and Glass Transition Phenomenology

In statistical mechanics the ergodicity hypothesis is very important as it states that
during the time evolution of a dynamical variable $A(t)$ it explores all the available
phase space. This implies that a time average is equivalent to an ensemble average
(performed with the statistical operator ρ). If a dynamical variable explores all the
phase space as the time goes to infinity it is called *ergodic*, and this is the case if and
only if [10]

$$\lim_{t \to \infty} \Phi_{AA}(t) \quad = \quad \lim_{z \to 0}[-z\Phi_{AA}(t)] = 0. \qquad (11.1)$$

where $\Phi_{AA}(t)$ is Kubo's relaxation function (7.36). For a classical dynamical
variable A the ergodicity condition (11.1) can be reformulated as

$$\lim_{t \to \infty} S_{AA}(t) \quad = \quad \lim_{z \to 0}[-zS_{AA}(t)] = 0. \qquad (11.2)$$

On the other hand, one can imagine situations in which a dynamical variable is
not able to explore the total phase space. Obviously the density variable $\rho_q(t)$ (7.6),
which describes the motion of the positions $R_\alpha(t)$ of the atoms becomes non-ergodic
if the liquid starts to *freeze*. As the phase transition towards the crystalline state
is a first-order one, usually a barrier (of nucleation processes) must be overcome
to reach this state once its free energy is lower than that of the liquid one. In
many situations this barrier is high enough that either the material cannot reach the
crystalline state or the experimentalist avoids crystallization by rapidly quenching.
In this situation the liquid can be undercooled below the crystallization temperature,
and can eventually also undergo a freezing transition, which is the *glass transition*
and leads to a frozen disordered structure. As the crystalline state is believed
to be that of lowest free energy the glassy state must be a metastable one. As
there are overwhelmingly many possibilities to form the glassy structure and the
corresponding free energy should be almost the same, this state is believed to be

W. Schirmacher, *Theory of Liquids and Other Disordered Media*, Lecture Notes
in Physics 887, DOI 10.1007/978-3-319-06950-0_11,
© Springer International Publishing Switzerland 2015

Fig. 11.1 "Free-energy landscape" of a glass in configuration space Ω_∞

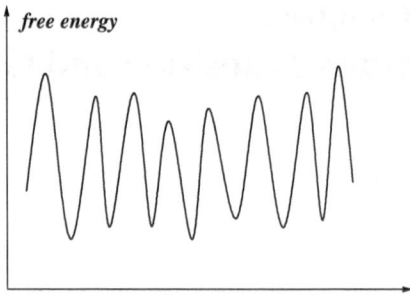

highly "quasi-degenerate", and it is believed to be composed to be formed of a "free-energy landscape" in configuration space (Fig. 11.1).

A convenient order parameter for monitoring structural arrest ($\Leftrightarrow$ "freezing") is the value the density relaxation function takes for $t \to \infty$:

$$f_q \equiv \lim_{t\to\infty} \frac{S(q,t)}{S(q,t=0)} = \lim_{t\to 0}[-z\frac{S(q,z)}{S(q,t=0)}] \qquad (11.3)$$

We can call the corresponding contribution to the density correlation function $S_\infty(q,t) \equiv S(q)f_q$ and have for the (classical) dynamical structure factor

$$S_\infty(q,\omega) = 2\pi S(q) f_q \delta(\omega) \qquad (11.4)$$

We see immediately that the *non-ergodicity parameter* f_q is the *Debye–Waller factor* of the glass. The presence of a "zero-phonon line" $\propto \delta(\omega)$ in the density fluctuation spectrum indicates the advent of structural arrest.

A non-zero non-ergodicity parameter f_q has also an important implication for the static density susceptibility $\chi(0)$. Inserting $\beta S_\infty(q,z) = \Phi_{q,\infty}(z) = -\beta S(q)f_q/z$ into (7.38) we obtain

$$\lim_{z\to 0} \chi(z) = \beta S(q)[1 - f_q] \qquad (11.5)$$

If (as we shall show to be the case) the glass transition is a *discontinuous* one, the static susceptibility $\chi(0)$ can therefore be expected to exhibit a jump from a "glass value" $\beta S(0)[1 - f_0]$ to a "liquid value" $\beta S(0) = \rho_0 \kappa_T$, where κ_T is the isothermal compressibility of the liquid (Fig. 11.2). At any finite frequency (and experiments are always effectively made at a finite frequency) this transition is blurred. In a real glass-forming liquid this transition is, however, not only blurred by the finite experimental time but also by solid-type activated molecular jump processes, which restores ergodicity at an exponentially large time. The corresponding activation energy is the low-temperature activation energy E_A of the viscosity, which has values of several eV per molecule.

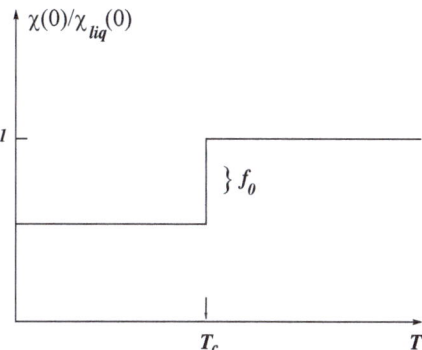

Fig. 11.2 Schematic behaviour of the static susceptibility at a (idealized) glass transition

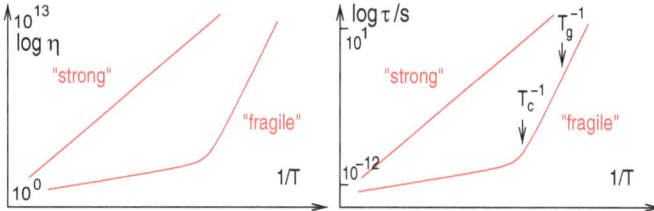

Fig. 11.3 *Left*: temperature dependence of the viscosity of a "strong" and a "fragile" glass-forming liquid. *Right*: temperature dependence of the relaxation time of a "strong" and "fragile" glass-forming liquid

In Fig. 11.3 we have plotted schematically the temperature dependence of the viscosity of supercooled liquids. According to Angell [1] one calls the glass-forming liquid "strong" if the viscosity exhibits an Arrhenius-type temperature dependence

$$\eta_S(T) = \eta_0 e^{-E_A/k_B T} \tag{11.6}$$

which gives a straight line in an Arrhenius-type graphical representation $\log \eta_S$ vs. $1/T$. If η_S in an Arrhenius plot shows a curved graph one calls the material "fragile". In this case it is very often possible to represent the temperature dependence of the viscosity by the phenomenological *Vogel–Fulcher–Tammann* formula

$$\eta_S(T) = \eta_0 e^{-E_B/k_B[T-T_0]} \tag{11.7}$$

where E_B and T_0 are adjustable parameters. Of course one can only use this formula for temperatures $T > T_0$. The cross-over to the steep Arrhenius behavior of fragile glasses in the glassy temperature regime is not described by this equation.

In the glass community the glass transition temperature T_g is defined in a sloppy way as that temperature at which the material does not flow any more, i.e., at which the viscosity exceeds some 10^{12} Ps.

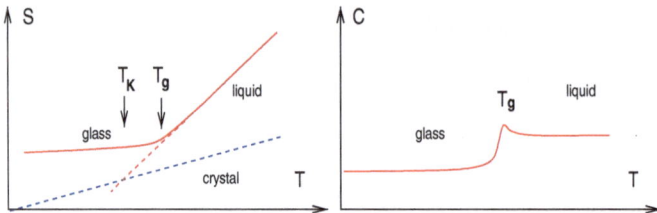

Fig. 11.4 *Left*: schematic sketch of the temperature dependence of the entropy of a supercooled liquid and the corresponding crystal. *Right*: schematic representation of the behaviour of the specific heat at the glass transition

In the glass-forming temperature regime one finds that, although the viscosity might be still finite, the material reacts to sudden external perturbations elastically. Following J.C. Maxwell one calls such a behavior visco-elastic behavior. The elastic shear modulus corresponding to rapidly time dependent perturbations is called G_∞ and Maxwell's relaxation time is defined as

$$\tau(T) = \eta_S(T)/G_\infty \qquad (11.8)$$

For times $t < \tau$ the material acts as a liquid, for times $t > \tau$ as a solid. If τ is much larger than a minute or so, the material is for all practical purposes a solid.[1] The glass transition temperature T_g corresponds to a relaxation time of an hour or so. It has to be emphasized that this definition of a glass transition temperature is an operational one and has nothing to do with a thermodynamic phase transition temperature. In the next section we shall call the temperature at which in an idealized glass transition the non-ergodicity parameter becomes non-zero the critical temperature T_c. We shall, however, see that this transition is very different from a thermodynamic phase transition. T_c is found to be some 20° higher than T_g in many glass-forming liquids.

If the cooling or heating rate of a differential scanning calorimetry (DSC) apparatus exceeds the inverse of the relaxation rate one usually observes a kink and even a maximum of the specific heat (Fig. 11.4). The corresponding "calorimetric glass transition temperature" is also called T_g and depends on the cooling rate. (Both definitions usually agree to each other to within a few degrees.) The entropy S, which is related to the specific heat by $C \propto dS/dT$ may then have the temperature dependence as depicted in Fig. 11.4. The right branch of the curve is the equilibrium branch, the left one is the out-of equilibrium one, which is history dependent. One speculates that for infinitely slow cooling the temperature dependence of S might follow the extrapolated equilibrium curve and might eventually cross the curve that corresponds to the crystalline state. As the perfect (insulating) crystal

---

[1]In this connection some people say that windows of old churches have been found to be thicker at the bottom than at the top and ascribe this to the finite viscosity of the window glass. This is, however, a fairy tale. The relaxation time of window glass at room temperature is orders of magnitudes larger than the age of the churches.

has only vibrational entropy and no configurational frozen-in disorder, its entropy must be always smaller than that of the glass, so that something must happen if the extrapolated liquid entropy approaches the crystalline one. This temperature is called *Kauzmann temperature* T_K and is smaller than T_g.

Recent theories of the glass transition, which focus on the regime below T_c [3, 15], have come up with a model of a thermodynamic-like ideal glass transition near T_K, but this approach is questioned by models [4, 5], in which the structural arrest below T_c occurs via kinetic constraints [16].

11.2   Idealized Glass Transition as Described by Mode-Coupling Theory

In the following we shall study the *normalized* density-density correlation function

$$\phi(q,t) = S(q,t)/S(q) \tag{11.9}$$

which, of course, obeys the same equation of motion (8.27) as $S(q,t)$

$$\frac{d^2}{dt^2}\phi(\mathbf{q},t) + \int_0^t d\tau M(\mathbf{q},\tau)\dot{\phi}(\mathbf{q},t-\tau) + \Omega_1^2\phi(\mathbf{q},t) = 0 \tag{11.10}$$

with

$$M(q,t) = \Omega_1(q)^2 m_q(t) = \Omega_q^2 \frac{1}{2V} \sum_{\substack{q_1+q_2=q \\ q+q_1+q_2=0}}^{q_1+q_2=q} V(\mathbf{q},\mathbf{q}_1,\mathbf{q}_2)S(q_1,t)S(q_2,t) \tag{11.11}$$

subject to the initial conditions

$$\phi(q,0) = 1; \qquad \dot{\phi}(q,0) = 0. \tag{11.12}$$

The remarkable feature of these equations is that they not only give a fair description of the collective dynamics of simple liquids, as we have seen in Sect. 8.4, but, beyond a critical value of the packing fraction $\eta_c \approx 0.52$ the solution $\phi(q,t)$ does no more decay towards 0 but towards a finite value f_q. This transition is called the idealized glass transition, and the corresponding temperature the idealized glass transition temperature T_c. The remarkable feature of this transition is that the structure fracture $S(q)$ changes smoothly with increasing η while going through the glass transition at $\eta = \eta_c$.

On the liquid side this implies the vanishing of the particle diffusion coefficient and the inverse of the viscosity according to a power-law $D \propto 1/\eta \propto [T-T_c]^\gamma$ with $\gamma \approx 2.6$ which fits measured viscosity and diffusivity data in the liquid range quite well. However, the most interesting feature of the solutions of the mode-coupling

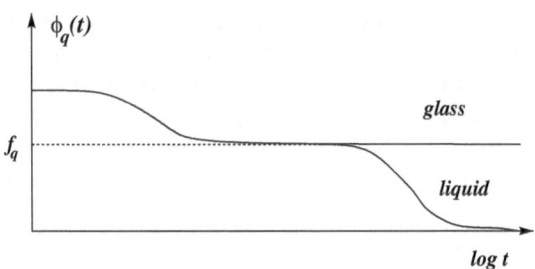

Fig. 11.5 Typical behavior of $\phi(q,t) = S(q,t)/S(q)$ near the idealized glass transition

equations are their highly nontrivial asymptotic long-time behaviour including two different scaling laws. It turns out that this behaviour is universal in a sense that its characteristic features do not depend on the details of the coupling vertex, but the predicted (and measured) critical exponents are non-universal, i.e., they slightly depend on the control parameters.

Near the glass transition, on the liquid side the function $\phi(q,t)$ develops a characteristic plateau as depicted in Fig. 11.5. In our present lecture we shall not discuss these features in terms of the full mode-coupling equations but by means of simple schematic equations, which, however exhibit the same asymptotics.

11.3 Phenomenological Mode-Coupling Theory and Schematic Model

11.3.1 Phenomenological Mode-Coupling Theory

From the foregoing it is clear that the derivation of the mode-coupling equations (11.11) together with (11.10) applies for simple hard-sphere-like one-component liquids. It has been tested against computer simulations of such liquids as well as experiments on hard-sphere colloid solutions [7].

However, many experimental data on quite more complicated glass-forming liquids like ortho-terphenyle, Potassium-Calcium Nitrate or Glycerol show the same critical behavior as predicted by the original mode-coupling equations. Götze [8] therefore proposed the following generalized phenomenological mode-coupling theory (MCT):

$$m_q(t) = \mathcal{F}_q\{\vec{V}, \phi(k,t)\} \tag{11.13}$$

with

$$\mathcal{F}_q\{\vec{V}, x_k\} = \sum_{m=1}^{m_0} \frac{1}{m!} \sum_{k_1 \cdots k_m} V^{(m)}(q, k_1 \cdots k_m) x_{k_1} \cdots x_{k_m} \tag{11.14}$$

Fig. 11.6 A fold singularity
as defined in catastrophe
theory

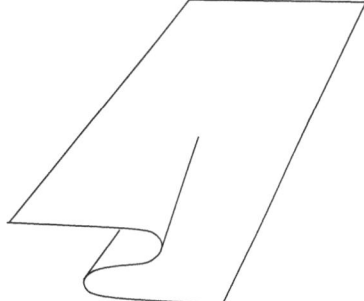

i.e., we have

$$m_q(t) = \sum_k V^{(1)}(q,k)\phi(k,t) + \frac{1}{2} \sum_{k_1,k_2} V^{(2)}(q,k_1,k_2)\phi(k_1,t)\phi(k_2,t) + \cdots$$

$$(11.15)$$

The search for non-ergodic solutions with $\phi(q,t \to \infty) = f_q \neq 0$ within this generalized set of mode-coupling equations can be shown to be equivalent to the search of topological singularities in the parameter space of the coefficients $V^{(m)}(q,k_1 \cdots k_m)$. Such singularities are the subject of the mathematical discipline "catastrophe theory", or, more modestly, the theory of bifurcations, developed by Thom [21] and Arnol'd [2]. The singularities relevant for the MCT glass transition are of the simplest type, namely the *fold* singularity or A_2 singularity, which can be visualized by crumpling a piece of paper (Fig. 11.6). Near the singularity the dynamics is dominated by a single eigenvalue of the stability matrix ("reduction theorem"), from which follows that the critical fluctuations are governed by a single function $G(t)$, and the wavevector q appears only in a prefactor ("factorization theorem" [6,8])

$$\phi(q,t) = f_q^c + h_q G(t) \tag{11.16}$$

11.3.2 Schematic Model

Instead of going into more detail of these considerations we study the bifurcation scenario with the following schematic model ("F_{12} model") for a q-independent correlation function $\phi(t)$ (with boundary condition $\phi(0) = 1$ and $\dot{\phi}(0) = 0$)

$$\frac{d^2}{dt^2}\phi(t) + v_0 \frac{d}{dt}\phi(t) + \Omega \int_0^t d\tau\, m(t-\tau) \frac{d}{d\tau}\phi(\tau) + \Omega^2 \phi(t) = 0, \tag{11.17}$$

Fig. 11.7 The function
$\lambda_1(f) = \frac{1}{1-f} - \lambda_2 f$

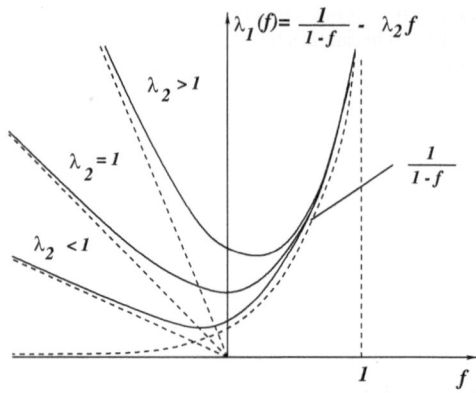

with the following memory function

$$m(t) = \lambda_1 \phi(t) + \lambda_2 \phi(t)^2 \tag{11.18}$$

(11.17) and (11.18) can be reformulated for the Laplace transforms as follows:

$$\frac{\phi(z)}{1 + z\phi(z)} - \frac{z + i v_0}{\Omega^2} = m(z) = \lambda_1 \phi(z) + \lambda_2 LT[\phi(t)^2]_z \tag{11.19}$$

We are now looking for non-ergodic asymptotic solutions $f \equiv \phi(t \to \infty) = -z\phi(z)|_{z\to 0}$ of (11.19). Such solutions must obey the equation

$$\frac{f}{1 - f} = \lambda_1 f + \lambda_2 f^2 \tag{11.20}$$

(the $z + i v_0$ term becomes negligible in comparison with $\frac{f}{z}$).

We observe immediately that $f \equiv 0$ is always a solution of this equation. However it can be shown that if there are several solutions to the mode-coupling equations, it is always the largest one which will be taken by the physical system (and also by the mathematical iteration). This implies also that f cannot be smaller than 0, so that we are looking for nonergodic solutions with $f > 0$ which are the solution of

$$\frac{1}{1 - f} = \lambda_1 + \lambda_2 f \tag{11.21}$$

If we inspect Fig. 11.7 in which the function $\lambda_1(f) = \frac{1}{1-f} - \lambda_2 f$ is plotted for different values of λ_2 we see that for $\lambda_2 < 1$ the minimum is situated at negative values of f, whereas for $\lambda_2 > 1$ the minimum is in the positive f regime. If we now increase λ_1 for a certain fixed value of λ_2 and look for the largest value of f at a given pair (λ_1, λ_2) we see that for $\lambda_2 < 1$ there is a *continuous* transition to a

Fig. 11.8 Phase diagram of
the F12 model

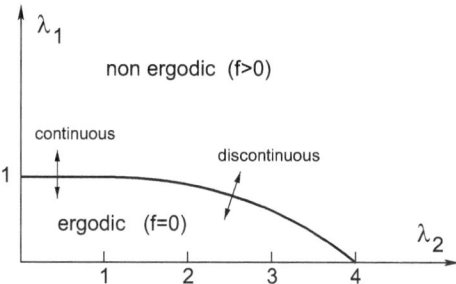

nonergodic state (*"Type A transition"*) whereas for $\lambda_2 > 1$ we have a *discontinuous* one (*"Type B transition"*).

In Fig. 11.8 we show the phase diagram in the $\lambda_1 - \lambda_2$ parameter plane. The type-A transition line is just given by $\lambda_1(\lambda_2) = 1$. The type-B line is given by $\lambda_1(\lambda_2) = \lambda_1(f_{min}) = 2\sqrt{\lambda_2} - \lambda_2$.

We would now like to study the dynamics right on the B-type transition line. In order to do so we divide the correlation function up as follows:

$$\phi(z) = f_c + hG(z) \tag{11.22}$$

and treat $|zG(z)|$ as a small parameter. Expanding both sides of (11.19) w.r. to $|zG(z)|$, and setting $\lambda \equiv 1 - f_c = \sqrt{1/\lambda_2}$ we obtain the following equation of motion:

$$zG^2(z) + \lambda LT[G(t)^2]_z + \lambda^3(z + iv_0) = 0 \tag{11.23}$$

We make now the ansatz

$$G(t) = A(t/t_0)^{-x} \quad \Leftrightarrow \quad G(z) = -\frac{1}{z}\Gamma(1-x)(-izt_0)^x \tag{11.24}$$

The right-hand side follows from (B.5) of the Appendix. We have also

$$LT[G(t)^2]_z = -\frac{1}{z}\Gamma(1-2x)(-izt_0)^{2x} \tag{11.25}$$

We see that for $z \to 0$ the "regular term" in (11.23) $\lambda^3(z + iv_0)$ can be neglected as $z \to 0$. Then (11.24) provides an asymptotic solution of (11.23) provided

$$\lambda \equiv \lambda(x) = \Gamma(1-x)^2/\Gamma(1-2x) \tag{11.26}$$

It is important to note that the solution—once the regular terms can be neglected—is not affected by changing the time scale t_0 (scale invariance).

Fig. 11.9 The function $\lambda(x)$

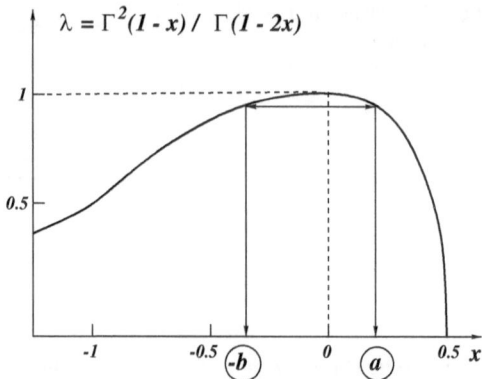

In Fig. 11.9 we have plotted the function $\lambda(x)$ and see that if we require $x \equiv a > 0$ and $\lambda > 0$ the value of a must be less than 0.5.

The "critical" relaxation law

$$\phi(t) = f_c + A(t/t_0)^{-a} \tag{11.27}$$

is called "β relaxation" and holds on both sides of the (idealized) glass transition.

Now we want to study the dynamics a little bit away from the critical line. In order to do so we define a "separation parameter" σ, which measures the distance from the critical line

$$\sigma \propto |\lambda_1 - \lambda_{1,c}| \quad \text{or} \quad \sigma \propto |T - T_c| \tag{11.28}$$

and we have now the equation (neglecting the "regular term")

$$zG^2(z) + \lambda LT[G(t)^2]_z = \frac{1}{z}\lambda^4(1 - \lambda)\sigma \tag{11.29}$$

We now are asking at which time scale the system realizes that it is away from the critical line and at which one it doesn't. The characteristic time which separates these regimes is obviously

$$t_\sigma = \frac{1}{\omega_\sigma} = \frac{\tilde{t}_0}{|\sigma|^{\frac{1}{2a}}} \tag{11.30}$$

where we now have defined a "microscopic" time scale $\tilde{t}_0$ which must be actually fixed by solving the MC equations numerically (which is most efficiently done in the time domain, solving (11.17) iteratively). As in the critical regime the time scale doesn't matter we can formulate the *first scaling law* or *β scaling law*

$$G(t) = c_\sigma g(t\omega_\sigma) \quad c_\sigma = \sqrt{|\sigma|}, \tag{11.31}$$

if first $\sigma \to 0$ and then $t \to \infty$ one has

$$g(t) \propto t^{-a} \tag{11.32}$$

Let us now consider the liquid phase in which we know that eventually the function $\phi(t)$ drops from the vicinity of f to 0. If we consider the vicinity of f by Eq. (11.29) there must be a dynamics leading away from f, which means that the z dependence of $G(z)$ must be stronger than $1/z$ for $z \to 0$. Such a solution exists (see Fig. 11.9) and introduces the second critical exponent $x = -b$ and we have

$$G(t) \propto -(t/\tau)^b \tag{11.33}$$

where the time scale τ is that at which the system realizes that it is not at criticality but in the ergodic liquid state. It turns out that the time scales τ and t_σ separate strongly from each other approaching the critical line ($\sigma \to 0$):

$$t_\sigma/\tau |\sigma|^{\frac{1}{2b}} \tag{11.34}$$

which implies

$$1/\tau = \frac{1}{t_0}|\sigma|^\gamma \qquad \gamma = \frac{1}{2a} + \frac{1}{2b}. \tag{11.35}$$

Note that the two critical exponents are related by (11.25), i.e.,

$$\lambda = \Gamma(1-a)^2/\Gamma(1-2a) = \Gamma(1+b)^2/\Gamma(1+2b) \tag{11.36}$$

The critical law (11.35) is that corresponding to structural relaxation, i.e., the diffusivity and the inverse viscosity go to zero accordingly.

It turns out that the entire remaining time dependence is governed by the scale τ which is the α *relaxation scale* and that the scaling function outside the $-t^b$ regime can be well approximated by a stretched exponential

$$\phi(t) \propto e^{-(t/\tau)^\beta} \tag{11.37}$$

where the exponent β must be determined numerically and has values near 0.5.

11.3.3 Summary of Anomalous Features Predicted by MCT

As stated in the beginning all the asymptotic properties of the schematic mode-coupling equations near the glass instability are the same as that of the full mode-coupling equations (11.10), (11.11) or (11.10), (11.15) by the reduction theorem.

Let us now summarize the salient features of the glass-transition scenario predicted by the mode-coupling equations:

- *factorization*

$$\phi(q,t) = f_q^c + h_q G(t)$$

- *β scaling*

$$G(t) \propto g(t/t_\sigma) \qquad t_\sigma = \frac{1}{\omega_\sigma} = \frac{\tilde{t}_0}{|\sigma|^{\frac{1}{2a}}} \qquad \sigma = T - T_c$$

$$\leftrightarrow$$

$$\frac{\chi''(\omega/\omega_\sigma)}{\chi''_{min}} = \begin{cases} (\omega/\omega_\sigma)^{-b} & \omega < \omega_\sigma \\ (\omega/\omega_\sigma)^a & \omega > \omega_\sigma \end{cases}$$

- *critical behavior of f_q and χ''_{min}*

$$f_q = f_q^c + h_q\sqrt{|\sigma|} \qquad \chi''_{min} \propto \sqrt{|\sigma|}$$

- *α scaling*

$$\phi(t) \propto e^{-(t/\tau)^\beta} \qquad \frac{1}{\tau} \propto D \propto \frac{1}{\eta} \propto \sigma^\gamma \qquad \gamma = \frac{1}{2a} + \frac{1}{2b}$$

$$\lambda = \Gamma(1-a)^2/\Gamma(1-2a) = \Gamma(1+b)^2/\Gamma(1+2b)$$

As stated above a large number of glass-forming materials exhibit this scenario.

11.4 Harmonic Vibrational Dynamics in Glasses

The anomalous vibrational features of glasses [9] cover a huge amount of literature. Here we just want to demonstrate that the glassy state with its frozen-in disorder produces some unusual features. On the other hand we would like to demonstrate once more the usefulness of the CPA.

11.4.1 Disordered Cubic Lattice and the Boson Peak

We consider again the model (6.80) on a simple cubic lattice [18]:

$$\frac{d^2}{dt^2} u_i(t) = -\sum_{j \neq i} K_{ij}[u_i(t) - u_j(t)] \equiv \sum_j D_{ij} u_j(t), \qquad (11.38)$$

The equation of motion of this system is

$$\frac{d^2}{dt^2} u_i(t) = -\sum_j K_{ij}(u_i(t) - u_j(t)) \qquad (11.39)$$

and we have the CPA equation with s being replaced by $-z^2$

$$\langle \frac{K - \Gamma(z)}{1 + (K - \Gamma(z))\frac{2}{Z\Gamma(z)}(1 + z^2 G_{ii}(z)} \rangle = 0 \qquad (11.40)$$

which can be reformulated as

$$\Gamma(z) = \langle \frac{K}{1 + (K - \Gamma(z))\frac{2}{Z\Gamma(z)}(1 + z^2 G_{ii}(z)} \rangle \qquad (11.41)$$

with the local Green's function given by (see (4.54)).

$$G(z) = \sum_{k \in BZ} \frac{1}{-z^2 + \Gamma(z) f(\mathbf{k})} \qquad (11.42)$$

The density of states is obtained from the Green's function according to

$$g(\omega) = 2\omega g(\omega^2) = -\frac{2\omega}{\pi} \Im m\{G(z)\} \qquad (11.43)$$

in Fig. 11.10 we have plotted the so-called reduced density of states $g(\omega)/\omega^2$ resulting from a numerical diagonalization of a model with a Gaussian $P(K)$ with

Fig. 11.10 Density of states, divided by ω^2 of a simple-cubic lattice with a Gaussian distribution of force constants with width parameter $\sigma/K_0 = 0.6$. The *full line* is a CPA calculation (solution of (11.41) and (11.42)), the *points* are the result of numerical diagonalizations of systems with periodic boundary conditions, averaged over different box sizes

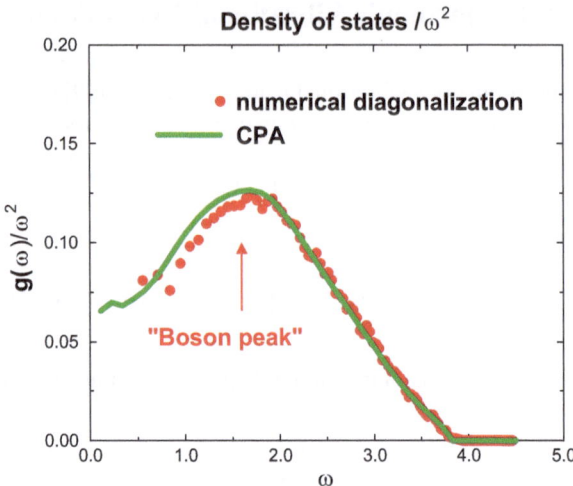

width-to-mean ratio $\sigma/K_0 = 0.6$ together with the CPA calculation. Clearly the CPA gives a good account of the data. It is also seen that there are no van-Hove singularities as in the ordered system (see the curve "$\sigma = 0$" in Fig. 11.11), but instead a maximum, which can be identified as the so-called boson peak.

We are now giving a physical explanation of the forming of such a peak in the reduced density of states of disordered solids. First we note that if we increase the width of the distribution $P(K)$ the system becomes unstable, because in the presence of negative values of K some atoms are now sitting on top of a potential hill instead at the bottom of a potential well. This instability manifests itself by the existence of negative eigenvalues ω_i^2 in the numerics, or, in the CPA, by the appearance of a portion of the density of states for negative values of ω^2. To have a "fine-tuning" of this instability we introduced a lower cutoff K_{min} in the Gaussian and put $\sigma = K_0$. The result is shown in Fig. 11.11.

It is seen that the more negative force constants we put in, the stronger becomes the boson peak. So it looks as if the boson peak is the precursor of the mentioned instability. This conclusion will be thoroughly corroborated in the rest of the present lectures.

11.4.2   Continuum CPA and Self-consistent Born Approximation, SCBA

We would like to simplify the CPA by the following steps:

(*i*) Getting rid of the unphysical cubic lattice by replacing the dispersion by a Debye law

Fig. 11.11 Density of states, divided by ω^2 of a simple-cubic lattice with a varying lower cutoff K_{min} and width parameter $\sigma/K_0 = 1$. The *full line* is a CPA calculation (solution of (11.41) and (11.42)), the *points* are the result of numerical diagonalizations of systems with periodic boundary conditions, averaged over different box sizes

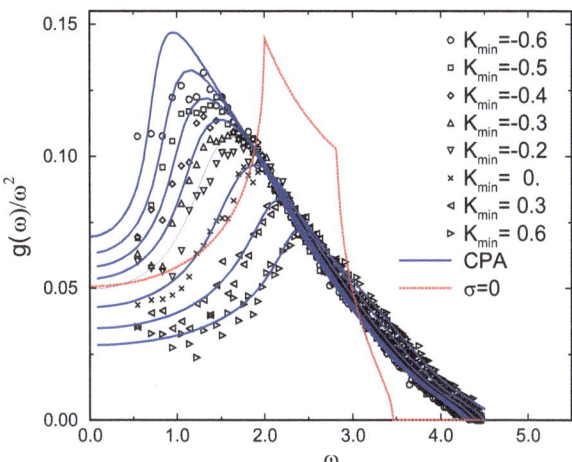

$$f(\mathbf{k}) \to a^2 k^2 \theta(k - k_D)$$

(ii) Expanding the CPA equation with respect to the fluctuations $K(\mathbf{r}) - \langle K \rangle$.

Step (i) leads to the continuum CPA equations, derived in Sect. 5.4 with the replacements $D \to \tilde{K} = a^2 K$ and $D(s) \to Q(z) = a^2 \Gamma(z)$:

$$0 = \left\langle \frac{\tilde{K}(\mathbf{r}) - Q(z)}{1 + \left(\tilde{K}(\mathbf{r}) - Q(z)\right)\frac{1}{3}\chi(z)} \right\rangle$$

$$\Longleftrightarrow$$

$$Q(z) = \left\langle \frac{\tilde{K}(\mathbf{r})}{1 + \left(\tilde{K}(\mathbf{r}) - Q(z)\right)\frac{1}{3}\chi(z)} \right\rangle \tag{11.44}$$

with

$$\chi(z) = \frac{1}{N} \sum_{|\mathbf{k}|<k_F} \frac{k^2}{-z^2 + Q(z)k^2} \tag{11.45}$$

In step (ii) we define

$$\tilde{K}(\mathbf{r}) = \tilde{K}_0 - \Delta\tilde{K}(\mathbf{r}) \qquad Q(s) = \tilde{K}_0 - \Sigma(z) \tag{11.46}$$

Now (11.44) and (11.45) take the form

$$0 = \left\langle \frac{\Delta\tilde{K}(\mathbf{r}) - \Sigma(z)}{1 - \left(\Delta\tilde{K}(\mathbf{r}) - \Sigma(z)\right)\frac{1}{3}\chi(z)} \right\rangle$$

Fig. 11.12 Density of states, divided by ω^2 for a generalized Debye model with fluctuating elastic constants $\tilde{K} \equiv c^2$ calculated for different disorder parameters $\gamma \propto \langle \delta \tilde{K}^2 \rangle / \langle \tilde{K} \rangle^2$, calculated in SCBA and CPA

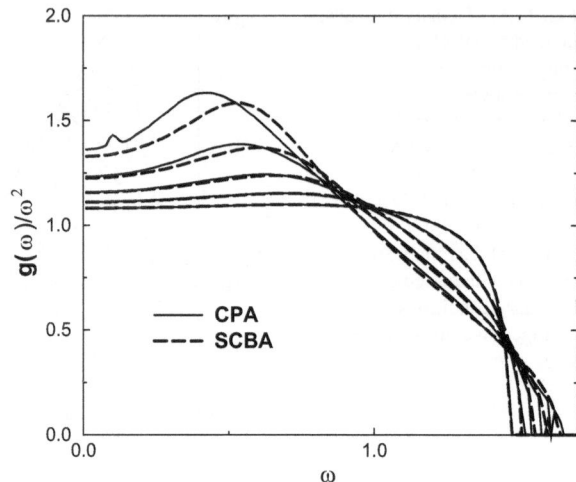

$$\Longleftrightarrow$$

$$\Sigma(z) = \left\langle \frac{\tilde{K}(\mathbf{r})}{1 - \left(\Delta \tilde{K}(\mathbf{r}) - \Sigma(z)\right)\frac{1}{3}\chi(z)} \right\rangle \tag{11.47}$$

with

$$\chi(z) = \frac{1}{N} \sum_{|\mathbf{k}|<k_F} \frac{k^2}{-z^2 + k^2[K_0 - \Sigma(z)]} \tag{11.48}$$

If we limit the expansion to second order in $\delta \tilde{K}$ and Σ we obtain the *self-consistent Born approximation, SCBA* [12]

$$\Sigma(z) = K_0^2 \gamma \chi(z) \tag{11.49}$$

with the disorder parameter $\gamma = \frac{1}{3K_0^2} \left\langle \left(\Delta K\right)^2 \right\rangle$

In Fig. 11.12 we compare the CPA for a Gaussian $P(K)$ with the SCBA with the same width parameters. As the results are not very different, one can use the SCBA for weakly fluctuating elastic constants[2] safely the SCBA instead of the CPA, especially if we don't want to compare our results to a simulation on a lattice. It is clear from Fig. 11.12 that the boson peak has nothing to do with a broadened van-Hove singularity as claimed sometimes in the literature [22], because all lattice-specific features have been removed from the theory. To understand the origin of the

---

[2]There exists a fully elaborated elasticity theory based on the SCBA with fluctuating shear modulus, which gives very similar result as the scalar model $K(\mathbf{r})$ [11, 17, 19].

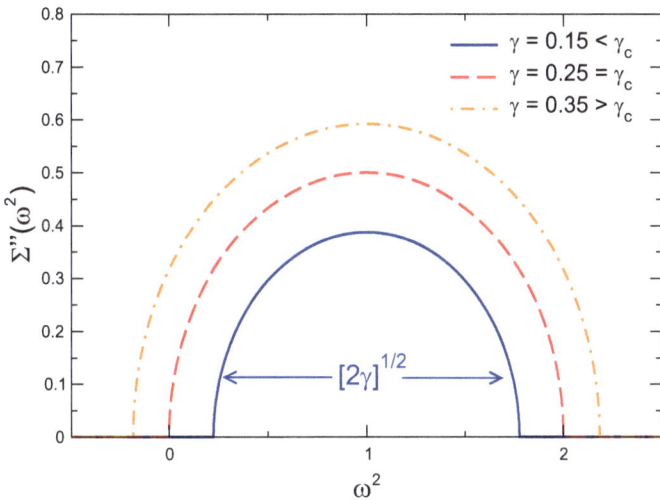

Fig. 11.13 Imaginary part of the "self energy" $\Sigma(\omega^2)$ for the simplified SCBA equation (11.50). If γ becomes larger than unity the system becomes unstable

boson peak we further simplify the SCBA equation (11.49) by replacing the sum over $\mathbf{k}$ by its dominant term at the maximum $|\mathbf{k}| = k_D$. If we use frequency units where $K_0 = k_D = 1$, we obtain

$$\Sigma(z) = 1 - Q(\tilde{z}) = \gamma \frac{1}{-z^2 + Q(\tilde{z})} \tag{11.50}$$

which is a quadratic equation for $Q(\tilde{z})$.

The solution of this equation yields a semicircular law in terms of the eigenvalues ω^2 for the imaginary part of the self energy Γ (Fig. 11.13) which is proportional to the enhancement of the "density of levels" $g(\omega^2) = g(\omega)$. If the variance-to mean square ratio γ ("disorder parameter") becomes comparable to unity, namely larger than $\gamma_c = 0.25$, the system becomes unstable [14]. If the disorder parameter is just a bit smaller than γ_c, a rapid rise of the spectrum occurs for $\omega^2 = 1 - 2\sqrt{\gamma}$, which in a plot $g(\omega)/\omega^2 = 2g(\omega^2)/\omega$ becomes a peak, which is actually the boson peak. The gap below this value occurs, because we omitted the $\mathbf{k}$ summation. If included, there is a crossover from the Debye behavior to the semicircular behavior. However, as one knows from random matrix theory [13, 20], the occurrence of a semicircular spectrum with width proportional to the mean of the distribution density of the matrix elements is a generic property of any random matrix.

So we can state that the boson peak marks the crossover from Debye's law (which is actually dictated by the translational symmetry of the equation of motion) to the semicircular law of the random Hamiltonian. The boson peak marks the crossover from wave-type vibrational excitations to disorder-dominated vibrations, characterized by a random Hamiltonian.

References

1. Angell, C.A.: Science **267**, 1924 (1995)
2. Arnol'd, V.: Catastrophe Theory. Addison-Wesley, Reading (1989)
3. Berthier, L., Biroli, G.: Rev. Mod. Phys. **83**, 587 (2011)
4. Biroli, G., Garrahan, J.P.: J. Chem. Phys. **138**, 12A301 (2013)
5. Chandler, D., Garrahan, J.P.: Ann. Rev. Phys. Chem. **61**, 191 (2010)
6. Götze, W.: In: Hansen, J.P., Levesque, D., Zinn-Justin, J. (eds.) Liquids, Freezing and the Glass Transition. Elsevier, Amsterdam. p. 287 (1991)
7. Götze, W.: Condens. Matter Phys. **124**, 873 (1998)
8. Götze, W.: Complex Dynamics of Glass-Forming Liquids: A Mode-Coupling Theory. Oxford University Press, Oxford (2008)
9. Kob, W., Binder, K.: Glassy Materials and Disordered Solids: An Introduction. World Scientific, London (2011)
10. Kubo, R.: J. Phys. Soc. Jpn. **12**, 570 (1957)
11. Marruzzo, A., Schirmacher, W., Fratalocchi, A., Ruocco, G.: Nat. Sci. Rep. **3**, 1407 (2013)
12. Maurer, E., Schirmacher, W.: J. Low Temp. Phys. **137**, 453 (2004)
13. Mehta, M.L.: Random Matrices. Elsevier, Amsterdam (2004)
14. Parisi, G.: J. Phys. Condens. Matter **15**, S765 (2003)
15. Parisi, G., Zamponi, F.: Rev. Mod. Phys. **82**, 789 (2010)
16. Ritort, F., Sollich, P.: Adv. Phys. **52**, 219 (2003)
17. Schirmacher, W.: Europhys. Lett. **73**, 892 (2006)
18. Schirmacher, W., Diezemann, G., Ganter, C.: Phys. Rev. Lett. **81**, 136 (1998)
19. Schirmacher, W., Ruocco, G., Scopigno, T.: Phys. Rev. Lett. **98**, 025501 (2007)
20. Stöckmann, H.-J.: Quantum Chaos: An Introduction. Cambridge University Press, Cambridge (1999)
21. Thom, R.: Structuras Stability and Morphogeneses: An Outline of a General Theory of Models. Addison-Wesley, Reading (1989)
22. Zorn, R.: Physics **4**, 44 (2011)

Chapter 12
Conclusions: Take-Home Messages

12.1 The Structure of Simple Liquids Is Essentially Determined by the Hard Core of the Potential

The prominent examples for simple liquids have been shown to be liquid metals. The structure factors of liquid metals are well described by a hard-sphere fluid, as demonstrated in Fig. 2.5. This means that the structure factor $S(q)$ (at nonzero wavenumber q) and its Fourier transform, the radial pair distribution function $g(r)$ are mainly determined by the hard-sphere-like core of the interatomic potential.

12.2 The Long-Wavelength Limit of the Structure Factors $S(0)$ Gives a Relation to the Thermal Properties of Soft Materials

In one-component liquids the long-wavelength limit of the structure factor is related to the isothermal compressibility (compressibility equation of state). In binary and multicomponent systems this statement remains true for the number fluctuations. The structure factor of the concentration fluctuations gives a relation to the stability of the multicomponent system.

12.3 The Perturbative Random-Phase Approximation (RPA) Describes Well the Deviations from the Hard-Sphere Structure

In contrast to the structure factors $S(q)$ at finite q their long-wavelength limit $S(0)$ is not only sensitive to the hard core of the interatomic potential but also to its longer-rage tail. This can be formally taken care of by the perturbative RPA. This

W. Schirmacher, *Theory of Liquids and Other Disordered Media*, Lecture Notes in Physics 887, DOI 10.1007/978-3-319-06950-0_12,
© Springer International Publishing Switzerland 2015

mean-field theory describes well the deviations of the compressibilities of simple liquids as calculated with the hard-sphere diameters taken from the finite-q structure factors. In liquid metals the predominantly repulsive nature of the interatomic potential leads to a relative hardening of the compressibility, whereas in liquids with (attractive) van-der-Waals forces one has a softening with respect to the hard-sphere compressibility.

12.4   In Binary Mixtures the Perturbative RPA Forms the Basis of the Regular-Solution Theory

The analog of the hard-sphere reference system of the number fluctuations is the non-interacting concentration-fluctuation structure factor in binary mixtures. This corresponds to the enthalpy of mixing of an ideal solution, derived from the ideal entropy of mixing. The perturbative RPA is equivalent to the Flory theory of regular solutions and—properly generalized—to the Flory theory of the thermodynamics of polymer solutions.

12.5   A Random Walk Is a Path of a Walker in Which the Direction Is Changed at Every Time Step Randomly and Is Described by the Diffusion Equation

Starting with a one-dimensional random walk using the combinatorial laws to go randomly to the left or right we derived the discrete master equation of random walk and from this in the continuum limit the one-dimensional diffusion equation. The two- and three-dimensional random walk and diffusion equation follow by superposition of the one-dimensional motion in the three Cartesian directions.

12.6   Fractals Have Non-integer Dimensionality

Fractals are self-similar objects, which can be ascribed a (Haussdorffian) dimensionality, which is smaller than the imbedding dimensionality. A prominent example is a percolating lattice, in which either sites or bonds are omitted randomly. At a certain site or bond concentration a phase transition from a connected to a disconnected object occurs. Right at the transition the percolation cluster (the cluster of remaining sides or bonds) is a fractal.

12.7   A Random Walk Is a Fractal with Fractal Dimension 2

Because the length ("time") along the random walk grows with the square of its spatial extension (mean-square displacement), its effective (i.e., fractal) dimensionality is 2. The corresponding fractal scaling properties become important for the structural properties of polymers.

12.8   The Thermodynamics of Polymers Are Governed by Their Fractal Scaling Properties

The de Gennes scaling theory of polymers is based on the fact that an idealized polymer is a random walk. The repulsive interaction, which is due to the non-intersecting property can be taken care of by adding a interaction term to the polymer's free energy. This changes the fractal dimension from 2 to 5/3.

12.9   The Dynamical Properties of Liquids Can Be Conventionally Described by Time Correlation Functions

The principal correlation function characterizing the dynamics of a liquid is van Hove's density-density correlation function $G(\mathbf{r}, t)$. Its time and space Fourier transform, the dynamical structure factor $S(q, \omega)$ can be measured by inelastic neutron or X-ray scattering. It is related with the dynamical response function of the density to an external potential perturbation by the fluctuation-dissipation theorem. Using a projection technique invented by Mori and Zwanzig a set of generalized Langevin equations for the van-Hove functions can be derived, which look like damped-harmonic oscillator equations, in which the damping constants is replaced by a memory functions.

12.10   The Collective Excitations of a Simple Liquid Can Be Well Described by Mode-Coupling Theory

By establishing a closure relation between the memory functions and the van-Hove functions a self-consistent theory (due to Götze) is obtained, which describes the collective sound-like excitations of liquid metals quite well. A remarkable property, which follows from this theory is the scaling with respect of the effective hard-sphere diameter is obeyed by the experimentally measured data.

12.11   Incoherent Liquid Dynamics Is Governed by Diffusion

Incoherent liquid dynamics, i.e., the individual-particle motion in a liquid is a modified random walk. The modifications can arise to step-wise motion (jump diffusion) and interactions, which can be described by a generalized random-phase approximation (RPA).

12.12   The Basic Polymer Dynamics Is Described by the Rouse Model, But a More Realistic Description Involves Disorder and Interaction

The dynamics of configurational changes occurs essentially along the chains. In the Rouse model it is described by the elastic vibrations along the changes, which are damped by the viscosity of the solvent. The inertial forces are neglected. More complicated models involve quenched disorder, activated motions and interactions via the response of the solvent (hydrodynamic forces).

12.13   The Liquid-to-Glass Transition Is a Transition from an Ergodic to a Nonergodic State

An ergodic system relaxes (by definition) towards its unique thermodynamic equilibrium state. In a non-ergodic state the system is caught inside a "pocket" of phase space from which it cannot escape. A liquid, frozen in a glassy structure, represents such a state. The dynamics near the transition involves algebraic time dependences and scaling laws, which are well described by a mean-field-like selfconsistent approximation (mode-coupling approximation).

12.14   Inside the Glassy State the High-Frequency Vibrations Show Irregularities Produced by the Quenched Disorder

The fact that on a microscopic scale the translational and rotational invariance are broken inside the glass leads to an irregular vibrational spectrum ("boson peak") in the THz range.

Appendix A
Fourier Transforms

In the present lecture notes the following convention for the Fourier transform is used[1]:

One-dimensional space

$$f(k) = \int dx\, e^{ikx} f(x) \qquad\qquad f(x) = \frac{1}{2\pi} \int dk\, e^{-ikx} f(k) \qquad (A.1)$$

Three-dimensional space

$$f(\mathbf{k}) = \int d^3r\, e^{i\mathbf{kr}} f(\mathbf{r}) \qquad\qquad f(\mathbf{r}) = \left(\frac{1}{2\pi}\right)^3 \int d^3k\, e^{-i\mathbf{kr}} f(\mathbf{k}) \qquad (A.2)$$

Case $f(\mathbf{r}) = f(|\mathbf{r}|) = f(r)$
We take $\mathbf{k} \parallel \mathbf{e}_z$

$$
\begin{aligned}
f(\mathbf{k}) &= \int_0^\pi \sin\theta d\theta \int_0^{2\pi} d\phi \int_0^\infty r^2 dr e^{ikr\cos\theta} f(r) \\
&= 2\pi \int_{-1}^1 d\cos\theta \int_0^\infty r^2 dr e^{ikr\cos\theta} f(r) \\
&= \frac{4\pi}{k} \int_0^\infty r dr \sin(kr) f(r) \qquad (A.3)
\end{aligned}
$$

Time

$$f(\omega) = \int dt\, e^{i\omega t} f(t) \qquad\qquad f(t) = \frac{1}{2\pi} \int d\omega\, e^{-i\omega t} f(\omega) \qquad (A.4)$$

---

[1]The boundaries of the integrals of the Fourier transforms are always $-\infty$ and $+\infty$.

W. Schirmacher, *Theory of Liquids and Other Disordered Media*, Lecture Notes in Physics 887, DOI 10.1007/978-3-319-06950-0,
© Springer International Publishing Switzerland 2015

Convolution theorem

$$h(\mathbf{r}) = \int d^3\tilde{\mathbf{r}} \, f(\mathbf{r} - \tilde{\mathbf{r}}) g(\tilde{\mathbf{r}})$$

$$= \int d^3\tilde{\mathbf{r}} \, f(\mathbf{r} - \tilde{\mathbf{r}}) \left(\frac{1}{2\pi}\right)^3 \int d^3\mathbf{k} \, e^{-i\mathbf{k}\tilde{\mathbf{r}}} g(\mathbf{k}) \quad \text{Substitution } \mathbf{r}' = \mathbf{r} - \tilde{\mathbf{r}}$$

$$= \left(\frac{1}{2\pi}\right)^3 \int d^3\mathbf{k} \, e^{-i\mathbf{k}\mathbf{r}} g(\mathbf{k}) \int d^3\mathbf{r}' \, e^{i\mathbf{k}\mathbf{r}'} f(\mathbf{r}')$$

$$= \left(\frac{1}{2\pi}\right)^3 \int d^3\mathbf{k} \, e^{-i\mathbf{k}\mathbf{r}} g(\mathbf{k}) f(\mathbf{k})$$

$$\Rightarrow \qquad h(\mathbf{k}) = \int d^3\mathbf{r} \, e^{i\mathbf{k}\mathbf{r}} h(\mathbf{r}) = f(\mathbf{k}) g(\mathbf{k}) \qquad\qquad\qquad (A.5)$$

Appendix B
Laplace Transforms

Common Laplace transform

$$f(s) = \int_0^\infty dt\, e^{-st} f(t) = \mathcal{L}[f(t)] \qquad \mathrm{Re}\{s\} > 0 \qquad \text{(B.1)}$$

Modified Laplace transform

$$f(z) = i \int_0^\infty dt\, e^{izt} f(t) = if(s = -iz) = LT[f(t)] \qquad \mathrm{Im}\{z\} > 0 \quad \text{(B.2)}$$

$$\mathcal{L}[\dot{f}(t)] = sf(s) - f(t = 0) \qquad LT[\dot{f}(t)] = -i\left(zf(z) + f(t=0)\right) \quad \text{(B.3)}$$

$$\mathcal{L}[\mathrm{const}] = \frac{\mathrm{const}}{s} \qquad LT[\mathrm{const}] = -\frac{\mathrm{const}}{z} \qquad \text{(B.4)}$$

$$\mathcal{L}[t^{\alpha-1}] = \frac{\Gamma(\alpha)}{s^\alpha} \qquad LT[t^{\alpha-1}] = i\frac{\Gamma(\alpha)}{(-iz)^\alpha} \qquad \text{(B.5)}$$

Convolution theorem for the Laplace transforms

$$h(t) = \int_0^t d\tau f(\tau)g(t-\tau) = \int_0^t d\tau g(\tau)f(t-\tau)$$

$$\mathcal{L}[h(t)] = f(s)g(s) \qquad LT[h(t)] = -if(z)g(z) \qquad \text{(B.6)}$$

$$\tilde{h}(t) = \int_0^t d\tau f(t-\tau)\dot{g}(\tau)$$

$$\mathcal{L}[h(t)] = f(s)\left(sg(s) - g(t=0)\right) \qquad LT[h(t)] = -f(z)\left(zg(z) + g(t=0)\right)$$

$$\text{(B.7)}$$

W. Schirmacher, *Theory of Liquids and Other Disordered Media*, Lecture Notes in Physics 887, DOI 10.1007/978-3-319-06950-0,
© Springer International Publishing Switzerland 2015

Hilbert–Stieltjes transform: We now derive the following important *properties of the (modified) Laplace transform:* Inserting for $f(t)$ in (B.2) the Fourier back transform in (A.4) we obtain

$$f(z) = \frac{1}{\pi} \int_{-\infty}^{\infty} d\bar{\omega} \frac{\frac{1}{2}f(\bar{\omega})}{\bar{\omega} - z} \tag{B.8}$$

which is the *Hilbert–Sijeltjes transform* of $\frac{1}{2}f(\bar{\omega})$. If we now insert the relation

$$\frac{1}{\bar{\omega} - \omega - i\epsilon} = \underbrace{\frac{\bar{\omega} - \omega}{(\bar{\omega} - \omega)^2 + \epsilon^2}}_{\to \ P(\frac{1}{\bar{\omega} - \omega})} + i \underbrace{\frac{\epsilon}{(\bar{\omega} - \omega)^2 + \epsilon^2}}_{\to \ \pi\delta(\bar{\omega} - \omega)} \tag{B.9}$$

Here $P\left(\frac{1}{x}\right)$ means the *principle part* of the singular function $1/x$

$$P\left(\frac{1}{x}\right) = \lim_{\epsilon \to 0} \frac{x}{x^2 + \epsilon^2} \tag{B.10}$$

we obtain from (B.8)

$$f(z = \omega + i\epsilon) \xrightarrow{\epsilon \to 0} P \int_{-\infty}^{\infty} \frac{d\omega}{\pi} \frac{\frac{1}{2}f(\bar{\omega})}{\bar{\omega} - \omega} + \frac{i}{2} f(\omega)$$
$$\equiv f'(\omega) + if''(\omega) \tag{B.11}$$

Taking these relations together we have

$$f(z = \omega + i\epsilon) = \frac{1}{\pi} \int_{-\infty}^{\infty} d\bar{\omega} \frac{f''(\bar{\omega})}{\bar{\omega} - z} = f'(\omega) + if''(\omega) \tag{B.12a}$$

$$f'(\omega) = \frac{1}{\pi} P \int_{-\infty}^{\infty} d\bar{\omega} \frac{f''(\bar{\omega})}{\bar{\omega} - \omega} \tag{B.12b}$$

$$f''(\omega) = \frac{1}{2} f(\omega) \tag{B.12c}$$

Appendix C
Velocity Autocorrelation, Diffusivity and Mean-Square Displacement

We would like to establish a connection between the *velocity autocorrelation function*

$$Z(t) = \langle v_x(t)v_x(0)\rangle = \langle v_y(t)v_y(0)\rangle = \langle v_z(t)v_z(0)\rangle \tag{C.1}$$

and the mean distance square walked by a random walker (mean-square displacement), who started at $\mathbf{r}(0)$ (on any medium, be it fractal or not)

$$
\begin{aligned}
\langle [\Delta\mathbf{r}(t)]^2\rangle &= \langle [\mathbf{r}(t) - \mathbf{r}(0)]^2\rangle \\
&= \langle [x(t) - x(0)]^2\rangle + \langle [y(t) - y(0)]^2\rangle + \langle [z(t) - z(0)]^2\rangle \\
&= \langle \Delta x(t)^2\rangle + \langle \Delta y(t)^2\rangle + \langle \Delta z(t)^2\rangle
\end{aligned} \tag{C.2}
$$

where $v_x(t) = \frac{\mathrm{d}}{\mathrm{d}t}x(t)$, from which follows

$$\Delta x(t) = \int_0^t \mathrm{d}\tau\, v_x(\tau). \tag{C.3}$$

We use the trivial identity

$$
\begin{aligned}
\int_0^t \mathrm{d}\tau\, \Delta x(\tau)v_x(\tau) &= -\int_0^t \mathrm{d}\tau\, v_x(\tau)\Delta x(\tau) + \Delta x(\tau)^2\Big|_0^t \\
\Rightarrow \qquad \Delta x(t)^2 &= 2\int_0^t \mathrm{d}\tau\, v_x(\tau)x(\tau) \\
&= 2\int_0^t \mathrm{d}\tau\, v_x(\tau)\int_0^\tau \mathrm{d}\tilde\tau\, v_x(\tilde\tau)
\end{aligned} \tag{C.4}
$$

W. Schirmacher, *Theory of Liquids and Other Disordered Media*, Lecture Notes in Physics 887, DOI 10.1007/978-3-319-06950-0,
© Springer International Publishing Switzerland 2015

so that we have

$$\left\langle \Delta x(t)^2 \right\rangle = 2 \int_0^t d\tau \int_0^\tau d\tilde{\tau} Z(\tau - \tilde{\tau}) = 2 \int_0^t d\tau \int_0^\tau d\tau' Z(\tau'), \qquad \text{(C.5)}$$

where we have made the substitution $\tau' = \tau - \tilde{\tau}$ in the second step. We now introduce the Laplace transform $Z(s)$ of $Z(t)$ and use the identity

$$\int_0^\infty dt e^{-st} \int_0^t d\tau f(\tau) = \frac{1}{s} f(s) \qquad \text{(C.6)}$$

to obtain

$$\Delta x^2(s) = \int_0^\infty e^{-st} dt \left\langle \Delta x(t)^2 \right\rangle = \frac{2}{s^2} Z(s) \qquad \text{(C.7)}$$

which finally yields

$$\Delta r^2(s) = \int_0^\infty e^{-st} dt \left\langle \Delta \mathbf{r}(t)^2 \right\rangle = \Delta x^2(s) + \Delta y^2(s) + \Delta z^2(s) = \frac{6}{s^2} Z(s) \quad \text{(C.8)}$$

We now apply the theorem

$$\lim_{s \to 0} s f(s) = \lim_{t \to \infty} f(t) \qquad \text{(C.9)}$$

to the function

$$f(t) = \frac{1}{6} \frac{d}{dt} \left\langle \Delta r(t)^2 \right\rangle. \qquad \text{(C.10)}$$

For the case of *normal diffusion* we obtain

$$D = \lim_{t \to \infty} \frac{1}{6} \frac{d}{dt} \left\langle \Delta r(t)^2 \right\rangle = \lim_{\epsilon \to 0} Z(s = \epsilon) = \lim_{\epsilon \to 0} \int_0^\infty dt e^{-\epsilon t} Z(t), \qquad \text{(C.11)}$$

which is the *Kubo relation* for the diffusivity. We now recall the *Nernst–Einstein relation* for the conductivity σ of charge carriers who perform a random walk

$$\sigma = e^2 \frac{n}{k_B T} D \qquad \text{(C.12)}$$

We can define a *generalized frequency dependent diffusivity* as

$$D(\omega) = \lim_{\epsilon \to 0} Z(s = i\omega + \epsilon) = \lim_{\epsilon \to 0} \int_0^\infty dt e^{i\omega t} e^{-\epsilon t} Z(t), \qquad \text{(C.13)}$$

where we have included an exponential factor with an infinitesimal (but positive) $\epsilon > 0$ for guaranteeing the convergence of the integral. $D(\omega)$ which is just the Laplace transform of $Z(t)$ with the complex Laplace frequency $s = \imath\omega + i\epsilon$.

The dynamic conductivity $\sigma(\omega)$ is related to $D(\omega)$ by the Nernst-Einstein relation

$$\sigma(\omega) = e^2 \frac{n}{k_B T} D(\omega) \qquad (C.14)$$